# Environmental Air Pollution

Causes, Effects and Control

# ENVIRONMENTAL AIR POLLUTION

## CAUSES, EFFECTS AND CONTROL

**P. N. Prasad**
**T. R. Amarnath**

CRESCENT
PUBLISHING CORPORATION

[*The responsibility for the facts stated, conclusions reached and plagiarism, if any, in this book is entirely that of the Author. And the Publisher bears no responsibility for them, whatsoever.*]

**Environmental Air Pollution**



First Published : 2010

ISBN 978-81-8342-179-9

*Published by* :

**CRESCENT PUBLISHING CORPORATION**
4831/24, Prahlad Street,
Ansari Road, Darya Ganj,
New Delhi - 110002
Ph. : 23272541, 23257835
Fax : 91- 011- 23257835
e-mail : campusbooks@hotmail.com

*Printed at :*
**Roshan Offset Printers**
Delhi

# Preface

Air pollution is the introduction of chemicals, particulate matter, or biological materials that cause harm or discomfort to humans or other living organisms, or damages the natural environment, into the atmosphere. Pollutants in the air which cause its contamination include natural emissions like ash from a volcanic eruption, but they mostly include the result of human activities such as the carbon monoxide gas from a motor vehicle exhaust or sulphur dioxide released from factories. These not only cause irreparable damage to the quality of the air that we breathe in, but also cause various health effects, which may range from subtle biochemical and physiological changes to difficulty in breathing, wheezing, coughing and aggravation of existing respiratory and cardiac conditions. Although efforts to control air pollution are numerous in their methods and control devices, the regulatory efforts have not yet proved satisfactory, since there needs to be much more of a concerted cooperation at the global scale. The book not only enumerates these causes, their effects and control mechanisms, but calls for a much more devoted and active effort to bring them to fruition.

It is hoped that this book, with its wide-ranging and in-depth content, critical insights, philosophical perspectives and detailed analysis of the issue, prove to be immensely resourceful to the readers.

**P. N. Prasad**
**T. R. Amarnath**

# Contents

# 1

# Introduction to Air Pollution

Air is the ocean we breathe. Air supplies us with oxygen which is essential for our bodies to live. Air is 99.9% nitrogen, oxygen, water vapor and inert gases. Human activities can release substances into the air, some of which can cause problems for humans, plants, and animals. The moment you step out of the house and are on the road you can actually see the air getting polluted; a cloud of smoke from the exhaust of a bus, car, or a scooter; smoke billowing from a factory chimney, flyash generated by thermal power plants, and speeding cars causing dust to rise from the roads. Natural phenomena such as the eruption of a volcano and even someone smoking a cigarette can also cause air pollution.

*The gaseous composition of unpolluted air*

| *The Gases* | *Parts per million (vol)* |
|---|---|
| Nitrogen | 756,500 |
| Oxygen | 202,900 |
| Water | 31,200 |
| Argon | 9,000 |
| Carbon Dioxide | 305 |
| Neon | 17.4 |
| Helium | 5.0 |
| Methane | 0.97-1.16 |
| Krypton | 0.97 |

| | |
|---|---|
| Nitrous oxide | 0.49 |
| Hydrogen | 0.49 |
| Xenon | 0.08 |
| Organic vapours ca. | 0.02 |

Air pollution is aggravated because of four developments: increasing traffic, growing cities, rapid economic development, and industrialisation. The Industrial Revolution in Europe in the 19th century saw the beginning of air pollution as we know it today, which has gradually become a global problem. There are several main types of pollution and well-known effects of pollution which are commonly discussed. These include smog, acid rain, the greenhouse effect, and "holes" in the ozone layer. Each of these problems has serious implications for our health and well-being as well as for the whole environment.

One type of air pollution is the release of particles into the air from burning fuel for energy. Diesel smoke is a good example of this particulate matter . The particles are very small pieces of matter measuring about 2.5 microns or about .0001 inches. This type of pollution is sometimes referred to as "black carbon" pollution. The exhaust from burning fuels in automobiles, homes, and industries is a major source of pollution in the air. Some authorities believe that even the burning of wood and charcoal in fireplaces and barbeques can release significant quanitites of soot into the air.

Another type of pollution is the release of noxious gases, such as sulfur dioxide, carbon monoxide, nitrogen oxides, and chemical vapors. These can take part in further chemical reactions once they are in the atmosphere, forming smog and acid rain.

Pollution also needs to be considered inside our homes, offices, and schools. Some of these pollutants can be created by indoor activities such as smoking and cooking. In the United States, we spend about 80-90% of our time inside buildings, and so our exposure to harmful indoor pollutants can be serious. It is therefore important to consider both indoor and outdoor air pollution.

## Causes of Air Pollution

Carbon dioxide is one the main pollutants that causes air pollution. This is because, although living beings do exhale carbon dioxide, this gas is harmful when emitted from other sources, which are caused due to human activity. An additional release of carbon dioxide happens due to various such activities. Carbon dioxide gas is used in various industries such as the oil industry and the chemical industry. The manufacturing process of most products would require the use of this gas. There are various human activities that add to the increased proportions of carbon dioxide in the atmosphere. The combustion of fossil fuels and the harmful effects of deforestation have all contributed towards the same. show that amongst the various gasses emitted during a volcanic eruption, carbon dioxide remains to be at least 40% of the emission. Scientists have now therefore identified carbon dioxide as one of those elements that have contributed to global warming.

Causes of air pollution are not limited to this. The combustion of fuels in automobiles, jet planes etc all cause the release of several primary pollutants into the air. The burning of fossil fuels in big cities which is seen at most factories, offices and even a large number of homes, it is no wonder that air pollution is increasing at an alarming rate. The release of other harmful gases all adds to the state that we see today. Although carbon dioxide plays an important role in various other processes like photosynthesis, breathing an excess of the same also causes harmful effects towards one's health.

The various causes of air pollution that releases harmful gases into the atmosphere are caused due to the increasing number of power plants and manufacturing units or industries that mostly have activities related to the burning of fuels. Besides, as mentioned earlier, most automobiles, marine vessels, activities that involve the burning of wood, fumes that are released from aerosol sprays, military activities that involve the use of nuclear weapons, all are the numerous causes of air pollution.

Carbon monoxide is another such gas which, although was present in the atmosphere earlier, is now considered to be a major pollutant. An excess of the same has a harmful effect on our system. There are many reasons why carbon monoxide can be released into the atmosphere as a result of human activities. This is also produced due to any fuel burning appliance and appliances such as gas water heaters, fireplaces, woodstoves, gas stoves, gas dryers, yard equipments as well as automobiles, which add to the increased proportion of this gas into the atmosphere.

Sulfur dioxide is yet another harmful pollutant that causes air pollution. Sulfur dioxide is emitted largely to the excessive burning of fossil fuels, petroleum refineries, chemical and coal burning power plants etc. Nitrogen dioxide when combined with sulfur dioxide can even cause a harmful reaction in the atmosphere that can cause acid rain.

Nitrogen dioxide is one more gas that is emitted into the atmosphere as a result of various human activities. An excess of nitrogen dioxide mainly happens due to most power plants seen in major cities, the burning of fuels due to various motor vehicles and other such sources, whether industrial or commercial that cause the increase in the levels of nitrogen dioxide.

These and a number of other hazardous air pollutants are emitted with the various numbers of activities that we carry out during the day which are the main causes of air pollution.

## Major Pollutants

Air pollution can cause health problems and it can also damage the environment and property. It has caused thinning of the protective ozone layer of the atmosphere, which is leading to climate change. Modernisation and progress have led to air getting more and more polluted over the years. Industries, vehicles, increase in the population, and urbanisation are some of the major factors responsible for air pollution. The following industries are among those that emit a great deal of pollutants into the air: thermal power plants, cement, steel, refineries, petro chemicals, and mines.

Air pollution results from a variety of causes, not all of which are within human control. Dust storms in desert areas and smoke from forest fires and grass fires contribute to chemical and particulate pollution of the air. The source of pollution may be in one country but the impact of pollution may be felt elsewhere.

The discovery of pesticides in Antarctica, where they have never been used, suggests the extent to which aerial transport can carry pollutants from one place to another. Probably the most important natural source of air pollution is volcanic activity, which at times pours great amounts of ash and toxic fumes into the atmosphere. The eruptions of such volcanoes as Krakatoa in Indonesia, Mt. St. Helens in Washington, USA and Katmai in Alaska, USA, have been related to measurable climatic changes

*Air Quality Standards*

| *Pollutants* | *Average Time* | *Concentration* |
|---|---|---|
| Sulphur dioxide ($SO_2$) | Annual average | 60 µg/m$^3$ |
| | 24 hour | 80 µg/m$^3$ |
| Oxides of Nitrogen ($NO_2$) | A.A | 60 µg /m$^3$ |
| | 24H | 80 µg /m$^3$ |
| Suspended Particulate Matter (SPM) | A.A | 140 µg/m$^3$ |
| | 24H | 200 µg/m$^3$ |
| Lead | A.A | 0.75 µg/m$^3$ |
| | 24H | 1.0 µg/m$^3$ |
| Carbon Monoxide | A.A | 2.0 µg/m$^3$ |
| | 24H | 4.0 µg/m$^3$ |
| Respirable Particulate Matter (RPM) | A.A | 60 µg/m$^3$ |
| | 24H | 100 µg/m$^3$ |

Listed below are the major air pollutants and their sources.

Carbon monoxide (CO)is a colourless, odourless gas that is produced by the incomplete burning of carbon-based fuels including petrol, diesel, and wood. It is also produced from the combustion of natural and synthetic products such as cigarettes. It lowers the amount of oxygen that enters our blood . It can slow our reflexes and make us confused and sleepy.

Carbon dioxide ($CO_2$) is the principle greenhouse gas emitted as a result of human activities such as the burning of coal, oil, and natural gases.

Chloroflorocarbons (CFC) are gases that are released mainly from air-conditioning systems and refrigeration. When released into the air, CFCs rise to the stratosphere, where they come in contact with few other gases, which leads to a reduction of the ozone layer that protects the earth from the harmful ultraviolet rays of the sun.

Leadis present in petrol, diesel, lead batteries, paints, hair dye products, etc. Lead affects children in particular. It can cause nervous system damage and digestive problems and, in some cases, cause cancer.

Ozone occur naturally in the upper layers of the atmosphere. This important gas shields the earth from the harmful ultraviolet rays of the sun. However, at the ground level, it is a pollutant with highly toxic effects. Vehicles and industries are the major source of ground-level ozone emissions. Ozone makes our eyes itch, burn, and water. It lowers our resistance to colds and pneumonia.

Nitrogen oxide ($No_x$) causes smog and acid rain. It is produced from burning fuels including petrol, diesel, and coal. Nitrogen oxides can make children susceptible to respiratory diseases in winters.

Suspended particulate matter (SPM) consists of solids in the air in the form of smoke, dust, and vapour that can remain suspended for extended periods and is also the main source of haze which reduces visibility. The finer of these particles, when breathed in can lodge in our lungs and cause lung damage and respiratory problems.

Sulphur dioxide ($SO_2$) is a gas produced from burning coal, mainly in thermal power plants. Some industrial processes, such as production of paper and smelting of metals, produce sulphur dioxide. It is a major contributor to smog and acid rain. Sulfur dioxide can lead to lung diseases.

**Outdoor Air Pollution**

Smog is a type of large-scale outdoor pollution. It is caused by chemical reactions between pollutants derived from different sources, primarily automobile exhaust and industrial emissions. Cities are often centers of these types of activities, and many suffer from the effects of smog, especially during the warm months of the year. Additional information about smog and its effects are available from Environment Canada and the Air Quality Management District (AQMD) in southern California.

For each city, the exact causes of pollution may be different. Depending on the geographical location, temperature, wind and weather factors, pollution is dispersed differently. However, sometimes this does not happen and the pollution can build up to dangerous levels. A temperature inversion occurs when air close to the earth is cooler than the air above it. Under these conditions the pollution cannot rise and be dispersed. Cities surrounded by mountains also experience trapping of pollution. Inversion can happen in any season. Winter inversions are likely to cause particulate and cabon monoxide pollution. Summer inversions are more likely to create smog.

Another consequence of outdoor air pollution is acid rain. When a pollutant, such as sulfuric acid combines with droplets of water in the air, the water (or snow) can become acidified. The effects of acid rain on the environment can be very serious. It damages plants by destroying their leaves, it poisons the soil, and it changes the chemistry of lakes and streams. Damage due to acid rain kills trees and harms animals, fish, and other wildlife. The U.S. Geological Survey (USGS), the Environmental Protection Agency (EPA), and Environment Canada are among the organisations that are actively studying the acid rain problem.

The Greenhouse Effect, also referred to as global warming, is generally believed to come from the build up of carbon dioxide gas in the atmosphere. Carbon dioxide is produced when fuels are burned. Plants convert carbon dioxide back to oxygen, but the release of carbon dioxide from human activities is higher than the world's plants can process. The situation is made worse since many of the earth's forests are being removed, and plant life is being damaged by acid rain. Thus, the amount of carbon dioxide in the air is continuing to increase. This buildup acts like a blanket and traps heat close to the surface of our earth. Changes of even a few degrees will affect us all through changes in the climate and even the possibility that the polar ice caps may melt.

Ozone depletion is another result of pollution. Chemicals released by our activities affect the stratosphere, one of the atmospheric layers surrounding earth. The ozone layer in the stratosphere protects the earth from harmful ultraviolet radiation from the sun. Release of chlorofluorocarbons (CFC's) from aerosol cans, cooling systems and refrigerator equipment removes some of the ozone, causing "holes"; to open up in this layer and allowing the radiation to reach the earth. Ultraviolet radiation is known to cause skin cancer and has damaging effects on plants and wildlife.

**Indoor Air Pollution**

Many people spend large portion of time indoors - as much as 80-90% of their lives. We work, study, eat, drink and sleep in enclosed environments where air circulation may be restricted. For these reasons, some experts feel that more people suffer from the effects of indoor air pollution than outdoor pollution.

There are many sources of indoor air pollution. Tobacco smoke, cooking and heating appliances, and vapors from building materials, paints, furniture, etc. cause pollution inside buildings. Radon is a natural radioactive gas released from the earth, and it can be found concentrated in basements in some parts of the United States.

Indoor air pollution refers to the physical, chemical, and biological characteristics of air in the indoor environment within a home, building, or an institution or commercial facility. Indoor air pollution is a concern in the developed countries, where energy efficiency improvements sometimes make houses relatively airtight, reducing ventilation and raising pollutant levels. Indoor air problems can be subtle and do not always produce easily recognised impacts on health. Different conditions are responsible for indoor air pollution in the rural areas and the urban areas.

In the developing countries, it is the rural areas that face the greatest threat from indoor pollution, where some 3.5 billion people continue to rely on traditional fuels such as firewood, charcoal, and cowdung for cooking and heating. Concentrations of indoor pollutants in households that burn traditional fuels are alarming. Burning such fuels produces large amount of smoke and other air pollutants in the confined space of the home, resulting in high exposure. Women and children are the groups most vulnerable as they spend more time indoors and are exposed to the smoke. In 1992, the World Bank designated indoor air pollution in the developing countries as one of the four most critical global environmental problems. Daily averages of pollutant level emitted indoors often exceed current WHO guidelines and acceptable levels. Although many hundreds of separate chemical agents have been identified in the smoke from biofuels, the four most serious pollutants are particulates, carbon monoxide, polycyclic organic matter, and formaldehyde. Unfortunately, little monitoring has been done in rural and poor urban indoor environments in a manner that is statistically rigorous.

In urban areas, exposure to indoor air pollution has increased due to a variety of reasons, including the construction of more tightly sealed buildings, reduced ventilation, the use of synthetic materials for building and furnishing and the use of chemical products, pesticides, and household care products. Indoor air pollution can begin within the building or be drawn in from outdoors. Other than nitrogen dioxide, carbon monoxide, and lead, there are a number of other pollutants that affect the air quality in an enclosed space.

Volatile organic compounds originate mainly from solvents and chemicals. The main indoor sources are perfumes, hair sprays, furniture polish, glues, air fresheners, moth repellents, wood preservatives, and many other products used in the house. The main health effect is the imitation of the eye, nose and throat. In more severe cases there may be headaches, nausea and loss of coordination. In the long term, some of the pollutants are suspected to damage to the liver and other parts of the body.

Tobacco smoke generates a wide range of harmful chemicals and is known to cause cancer. It is well known that passive smoking causes a wide range of problems to the passive smoker (the person who is in the same room with a smoker and is not himself/herself a smoker) ranging from burning eyes, nose, and throat irritation to cancer, bronchitis, severe asthma, and a decrease in lung function.

## Sources of Air Pollution

Air pollutants consist of gaseous pollutants, odours, and SPM, (suspended particulate matter) such as dust, fumes, mist, and smoke. The concentration of these in and near the urban areas causes severe pollution to the surroundings. The largest sources of human-created air pollution are energy generation, transportation, and industries that use a great deal of energy sources.

Depending on their source and interactions with other components of the air, they can have different chemical compositions and health impacts. Since these pollutants are generally concentrated in and around urban areas, the outdoor urban pollution levels are far higher than in the rural areas.

Fires are another major source of air pollution and can lead to severe problems if the smoke is inhaled for a period of time. These fires can either be forest fires, oil well fires, burning of leaves in the backyard or as in the case of rural areas, large-scale burning of agricultural waste. Other sources include industries and power plants located in these areas.

## Health Impacts of Air Pollution

Since the onset of the industrial revolution, there has been a steady change in the composition of the atmosphere mainly due to the combustion of fossil fuels used for the generation of energy and transportation.

Air pollution is a major environmental health problem affecting the developing and the developed countries alike. The effects of air pollution on health are very complex as there are many different sources and their individual effects vary from one to the other. It is not only the ambient air quality in the cities but also the indoor air quality in the rural and the urban areas that are causing concern. In fact in the developing world the highest air pollution exposures occur in the indoor environment.

Air pollutants that are inhaled have serious impact on human health affecting the lungs and the respiratory system; they are also taken up by the blood and pumped all round the body. These pollutants are also deposited on soil, plants, and in the water, further contributing to human exposure. As you read on you can learn about health impacts of specific air pollutants.

Indoor air pollution can be particularly hazardous to health as it is released in close proximity to people. It is stated that a pollutant released indoors is many times more likely to reach the lung than that released outdoors. In the developing countries a fairly large portion of the population is dependent on biomass for their energy requirements. These include wood, charcoal, agricultural residue, and animal waste. Open fires used for cooking and heating are commonly found in the household both in the rural and the urban areas. The stove is often at floor level, adding to the risk of accident and the hygiene factor. In addition, they are often not fitted with a chimney to remove the pollutants. In such households the children and women are most likely to be affected, as they are the group that spends more time indoors. The main pollutant in this environment is the SPM. In fact, death due to indoor air pollution, mainly particulate matters, in the rural areas of India are one of the highest in the world. Many of the deaths are due to acute respiratory infections in children; others are due to cardiovascular diseases, lung cancer, and chronic respiratory diseases in adults. If emissions are high and ventilation is poor, household use of coal and biomass can severely affect the indoor air quality.

Pollutant emissions per meal are also very high compared to those of other fuels. Household use of fossil fuel is also fairly common in the developing countries, particularly coal-both bituminous and lignite. These are particularly damaging as they burn inefficiently and emit considerable quantities of air pollutants. If emissions are high and ventilation poor, then the exposure levels to the gases emitted are far higher. The most harmful of the gases and agents that are emitted are particulate matter, carbon dioxide, polycyclic organic matter, and formaldehyde. The indoor concentrations of these are far higher than the acceptable levels and is cause for concern in rural areas.

Some of these gases can seriously and adversely affect the health of the population and should be given due attention by the concerned authority. The gases mentioned below are mainly outdoor air pollutants but some of them can and do occur indoor depending on the source and the circumstances.

— *Tobacco smoke.* Tobacco smoke generates a wide range of harmful chemicals and is a major cause of ill health, as it is known to cause cancer, not only to the smoker but affecting passive smokers too. It is well-known that smoking affects the passive smoker (the person who is in the vicinity of a smoker and is not himself/herself a smoker) ranging from burning sensation in the eyes or nose, and throat irritation, to cancer, bronchitis, severe asthma, and a decrease in lung function.

— *Biological pollutants.* These are mostly allergens that can cause asthma, hay fever, and other allergic diseases.

— *Volatile organic compounds.* Volatile compounds can cause irritation of the eye, nose and throat. In severe cases there may be headaches, nausea, and loss of coordination. In the longer run, some of them are suspected to cause damage to the liver and other parts of the body.

— *Formaldehyde.* Exposure causes irritation to the eyes, nose and may cause allergies in some people.

— *Lead.* Prolonged exposure can cause damage to the nervous system, digestive problems, and in some cases cause cancer. It is especially hazardous to small children.

- *Radon.* A radioactive gas that can accumulate inside the house, it originates from the rocks and soil under the house and its level is dominated by the outdoor air and also to some extent the other gases being emitted indoors. Exposure to this gas increases the risk of lung cancer.
- *Ozone.* Exposure to this gas makes our eyes itch, burn, and water and it has also been associated with increase in respiratory disorders such as asthma. It lowers our resistance to colds and pneumonia.
- *Oxides of nitrogen.* This gas can make children susceptible to respiratory diseases in the winters.
- *Carbon monoxide.* CO (carbon monoxide) combines with haemoglobin to lessen the amount of oxygen that enters our blood through our lungs. The binding with other haeme proteins causes changes in the function of the affected organs such as the brain and the cardiovascular system, and also the developing foetus. It can impair our concentration, slow our reflexes, and make us confused and sleepy.
- *Sulphur dioxide.* $SO_2$ (sulphur dioxide) in the air is caused due to the rise in combustion of fossil fuels. It can oxidise and form sulphuric acid mist. $SO_2$ in the air leads to diseases of the lung and other lung disorders such as wheezing and shortness of breath. Long-term effects are more difficult to ascertain as $SO_2$ exposure is often combined with that of SPM.
- *SPM (suspended particulate matter).* Suspended matter consists of dust, fumes, mist and smoke. The main chemical component of SPM that is of major concern is lead, others being nickel, arsenic, and those present in diesel exhaust. These particles when breathed in, lodge in our lung tissues and cause lung damage and respiratory problems. The importance of SPM as a major pollutant needs special emphasis

as a) it affects more people globally than any other pollutant on a continuing basis; b) there is more monitoring data available on this than any other pollutant; and c) more epidemiological evidence has been collected on the exposure to this than to any other pollutant.

Air pollution can affect our health in many ways with both short-term and long-term effects. Different groups of individuals are affected by air pollution in different ways. Some individuals are much more sensitive to pollutants than are others. Young children and elderly people often suffer more from the effects of air pollution. People with health problems such as asthma, heart and lung disease may also suffer more when the air is polluted. The extent to which an individual is harmed by air pollution usually depends on the total exposure to the damaging chemicals, i.e., the duration of exposure and the concentration of the chemicals must be taken into account.

Examples of short-term effects include irritation to the eyes, nose and throat, and upper respiratory infections such as bronchitis and pneumonia. Other symptoms can include headaches, nausea, and allergic reactions. Short-term air pollution can aggravate the medical conditions of individuals with asthma and emphysema. In the great "Smog Disaster" in London in 1952, four thousand people died in a few days due to the high concentrations of pollution.

Long-term health effects can include chronic respiratory disease, lung cancer, heart disease, and even damage to the brain, nerves, liver, or kidneys. Continual exposure to air pollution affects the lungs of growing children and may aggravate or complicate medical conditions in the elderly. It is estimated that half a million people die prematurely every year in the United States as a result of smoking cigarettes.

Research into the health effects of air pollution is ongoing. Medical conditions arising from air pollution can be very expensive. Healthcare costs, lost productivity in the workplace, and human welfare impacts cost billions of dollars each year.

## Smog and Air Pollution

The term smog was first used in 1905 by Dr H A Des Voeux to describe the conditions of fog that had soot or smoke in it. Smog is a combination of various gases with water vapour and dust. A large part of the gases that form smog is produced when fuels are burnt. Smog forms when heat and sunlight react with these gases and fine particles in the air. Smog can affect outlying suburbs and rural areas as well as big cities. Its occurrences are often linked to heavy traffic, high temperatures, and calm winds. During the winter, wind speeds are low and cause the smoke and fog to stagnate; hence pollution levels can increase near ground level. This keeps the pollution close to the ground, right where people are breathing. It hampers visibility and harms the environment. Heavy smog is greatly decreases ultraviolet radiation. In fact, in the early part of the 20th century, heavy smog in some parts of Europe resulted in a decrease in the production of natural vitamin D leading to a rise in the cases of rickets. Smog causes a misty haze similar to fog, but very different in composition. In fact the word smog has been coined from a combination of the words fog and smoke. Smog refers to hazy air that causes difficult breathing conditions.

The most harmful components of smog are ground-level ozone and fine airborne particles. Ground-level ozone forms when pollutants released from gasoline and diesel-powered vehicles and oil-based solvents react with heat and sunlight. It is harmful to humans, animals, and plants.

The industrial revolution in the 19th century saw the beginning of air pollution in Europe on a large scale and the presence of smog mainly in Britain. The industries and the households relied heavily on coal for heating and cooking. Due to the burning of coal for heat during the winter months, emissions of smoke and sulphur dioxide were much greater in urban areas than they were during the summer months. Smoke particles trapped in the fog gave it a yellow/black colour and this smog often settled over cities for many days.

The effects of smog on human health were evident, particularly when smog persisted for several days. Many people suffered respiratory problems and increased deaths were recorded, notably those relating to bronchial causes. A haze of dense harmful smog would often cover the city of London. The first smog-related deaths were recorded in London in 1873, when it killed 500 people. In 1880, the toll was 2000. London had one of its worst experiences with smog in December 1892. It lasted for three days and resulted in about 1000 deaths. London became quite notorious for its smog. By the end of the 19th century, many people visited London to see the fog. Despite gradual improvements in air quality during the 20th century, another major smog occurred in London in December 1952. The Great London Smog lasted for five days and resulted in about 4000 more deaths than usual. In response to the Great London Smog, the government passed its first Clean Air Act in 1956, which aimed to control domestic sources of smoke pollution by introducing smokeless zones. In addition, the introduction of cleaner coals led to a reduction in sulphur dioxide pollution. In the 1940s, severe smog began covering the city of Los Angeles in the USA.

Relatively little was done to control any type of pollution or to promote environmental protection until the middle of the 20th century. Today, smoke and sulphur dioxide pollution in cities is much lower than in the past, as a result of legislation to control pollution emissions and cleaner emission technology.

**Acid Aain**

Another effect of air pollution is acid rain. The phenomenon occurs when sulphur dioxide and nitrogen oxides from the burning of fossil fuels such as, petrol, diesel, and coal combine with water vapour in the atmosphere and fall as rain, snow or fog. These gases can also be emitted from natural sources like volcanoes. Acid rain causes extensive damage to water, forest, soil resources and even human health. Many lakes and streams have been contaminated and this has led to the disappearance of some species of fish in Europe, USA and Canada as also extensive damage to forests and other forms of life. It is said that it can corrode buildings and be hazardous to human health. Because the contaminants are carried long distances, the sources of acid rain are difficult to pinpoint and hence difficult to control. For example, the acid rain that may have damaged some forest in Canada could have originated in the industrial areas of USA. In fact, this has created disagreements between Canada and the United States and among European countries over the causes of and solutions to the problem of acid rain. The international scope of the problem has led to the signing of international agreements on the limitation of sulphur and nitrogen oxide emissions.

**Flyash**

With the boom in population and industrial growth, the need for power has increased manifold. Nearly 73% of India's total installed power generation capacity is thermal, of which 90% is coal-based generation, with diesel, wind, gas, and steam making up the rest. Thermal power generation through coal combustion produces minute particles of ash that causes serious environmental problems.

Commonly known as fly ash, these ash particles consist of silica, alumina, oxides of iron, calcium, and magnesium and toxic heavy metals like lead, arsenic, cobalt, and copper. The 80-odd utility thermal power stations in India use bituminous coal and produce large quantities of fly ash. According to one estimate, up to 150 million tonnes of fly ash will be produced in India in the year 2000, primarily by thermal power plants and, to a lesser extent, by cement and steel plants and railways. This poses problems in the form of land use, health hazards, and environmental dangers. Both in disposal and in utilisation utmost care has to be taken to safeguard the interest of human life, wild life, and such other considerations.

The prevalent practice is to dump fly ash on wastelands, and this has lain to waste thousands of hectares all over the country. To prevent the fly ash from getting airborne, the dumping sites have to be constantly kept wet by sprinkling water over the area. The coal industry in USA spends millions of dollars on lining fly ash dumping grounds. It lowers soil fertility and contaminates surface and ground water as it can leach into the subsoil. When fly ash gets into the natural draining system, it results in siltation and clogs the system. It also reduces the pH balance and portability of water. Fly ash interferes with the process of photosynthesis of aquatic plants and thus disturbs the food chain. Besides, fly ash corrodes exposed metallic structures in its vicinity.

In Delhi, the problem of fly ash is particularly severe as three power stations are located here. Being very minute, fly ash tends to remain airborne for a very long period leading to serious health problems as the airborne ash can enter the body. It causes irritation to eyes, skin, and nose, throat, and respiratory tract. Repeated inhalation of fly ash dust containing crystalline silica can cause bronchitis and lung cancer.

**Tackling the Problem of Flyash**

Fly ash management has taken considerable strides over the past few years. Researches have been attempting to convert this waste into wealth by exploring viable avenues for fly ash management. Fly ash is oxide-rich and can be used as the raw material for different industries.

Today, fly ash bricks can be used as a building material. The American Embassy in India has used fly ash bricks in some of its recent construction. Use of fly ash as a part replacement of cement in mortar and concrete has started with the Indian Institute of Technology, Delhi taking the lead. Use of fly ash in the construction of roads and embankments has been successfully demonstrated in the country and it is gaining acceptance. The NTPC (National Thermal Power Corporation) is setting up two fly ash brick manufacturing plants at Badarpur and Dadri near Delhi.

At TERI, researchers have proven that fly ash dumps can be reclaimed by suitable addition of organic matter and symbiotic fungi, making it commercially viable for activities like floriculture and silviculture. TERI researchers have successfully reclaimed a part of an ash pond at the Badarpur Thermal Power Station by introducing a mycorrhizal fungi-based organic bio-fertiliser. As the fungus germinates, it sustains on the partner plant and quickly spreads to the roots and beyond. It improves the plant's water and nutrient uptake, helps in the development of roots and soil-binding, stores carbohydrates and oils for use when needed, protects the plants from soil-borne diseases, and detoxifies contaminated soils. This helps in keeping both air and water pollution under control. It also helps revive wastelands and saves millions of litres of precious water from going down the fly ash slurries. Marigold, tuberose, gladiolus, carnation, sunflower, poplar, sheesham, and eucalyptus now grow at the demonstration site of the power station.

## Prevention of Air Pollution

In many countries in the world, steps are being taken to stop the damage to our environment from air pollution. Scientific groups study the damaging effects on plant, animal and human life. Legislative bodies write laws to control emissions. Educators in schools and universities teach students, beginning at very young ages, about the effects of air pollution.

The first step to solving air pollution is assessment . Researchers have investigated outdoor air pollution and have developed standards for measuring the type and amount of some serious air pollutants.

Once exposure levels have been set, steps can be undertaken to reduce exposure to air pollution. These can be accomplished by regulation of man-made pollution through legislation. Many countries have set controls on pollution emissions for transportation vehicles and industry. This is usually done to through a variety of coordinating agencies which monitor the air and the environment. At the United Nations, the Atmosphere Management Program carries out world wide environmental projects. In the United States, the primary federal agency is the Environmental Protection Agency. Many state and local organisations also participate in monitoring and controlling the environment. These include the San Francisco Bay Area's Air Quality Management District (B.A.A.Q.M.D.), the Air Quality

Management District in southern California, the Environmental Protection Agency of California, SmogBusters of southern California, and the Galveston-Houston Association for Smog Prevention (GHASP).

Prevention is another key to controlling air pollution. The regulatory agencies mentioned above play an essential role in reducing and preventing air pollution in the environment. In addition, it is possible to prevent many typès of air pollution that are not regulated through personal, careful attention to our interactions with the environment. In the United States, most household products come with instructions about safe use

Building materials should be reviewed for potential harmful effects. Adequate ventilation is also a key to controlling exposure to indoor air pollution. Home and work environments should be monitored for adequate air flow and proper exhaust systems installed.

One of the most dangerous air pollutants is cigarette smoke. Restricting smoking is an important key to a healthier environment. Legislation to control smoking is in effect in some locations, but personal exposure should be monitored and limited wherever possible. Only through the efforts of scientists, business leaders, legislators, and individuals can we reduce the amount of air pollution on the planet. This challenge must be met by all of us in order to assure that a healthy environment will exist for ourselves and our children.

# 2

# Indoor Air Pollution

In developing countries, the problem of indoor air pollution far outweighs the ambient air pollution. There are four principal sources of pollutants of indoor air: (i) combustion, (ii) building material, (iii) the ground under the building, and (iv) bioaerosols. In developed countries the most important indoor air pollutants are radon, asbestos, volatile organic compounds, pesticides, heavy metals, animal dander, mites, moulds and environmental tobacco smoke. However, in developing countries the most important indoor air pollutants are the combustion products of unprocessed solid biomass fuels used by the poor urban and rural folk for cooking and heating.

Approximately half the world's population and up to 90% of rural households in developing countries still rely on unprocessed biomass fuels such as wood, dung and crop residues. A recent report of the World Health Organisation (WHO) asserts the rule of 1000 which states that a pollutant released indoors is one thousand times more likely to reach people's lung than a pollutant released outdoors. It has been estimated that about half a million women and children die each year from indoor air pollution in India. Compared to other countries, India has among the largest burden of disease due to the use of dirty household fuels and 28% of all deaths due to indoor air pollution in developing countries occur in India.

The type of fuels used by a household is determined maily by its economic status. In the energy ladder, biomass fuels namely animal dung, crop residues and wood, which are the dirtiest fuels, lie at the bottom and are used mostly by very poor people. Electricity, which is the most expensive, lies at the top of ladder and it is also the cleanest fuel. The 1991 National Census for the first time inquired about the fuel used for coking. It revealed that about 90% of the rural population relied upon the biomass fuels like animal dung, crop residues and wood. A small portion used coal. Nation-wide about 78% of the population relied upon the biomass fuels and 3% on coal.

## Air Pollutants from Biomass Combustion

It has been estimated that more than half world's households cook their food on the unprocessed solid fuels that typically release at least 50 times more noxious pollutants than gas. The stoves or *chullah* used for cooking are not energy efficient. The fuels re not burned completely. The incomplete combustion of biomass releases complex mixture of organic compounds, which include suspended particulate matter, carbon monoxide, poly organic material (POM), poly aromatic hydrocarbons (PAH), formaldehyde, *etc.* The biomass may also contain intrinsic contaminants such as sulphur, trace metals, *etc.*

### Particulates

In recent years a large number of studies of health impact of suspended particulate air pollution have been undertaken in developing countriesThese studies show remarkable consistency in the relationship observed between changes in daily ambient suspended particulate levels and changes in mortality. Smithestimated the health risk from exposure to particulate air pollution by applying the mean risk per unit ambient concentrations based on the results of some urban epidemiological studies. The range of risk was found to be 1.2 - 4.4% increased mortality per 10 mg/m$^3$ incremental increase in concentration of respirable suspended particles (PM). For the calculations of estimates, it was assumed that the health risk has linear relationship to exposure, the risk factors determined for urban centers of developed nations were used as standards; where the PMdata were not available, 50% of suspended particulate matter (SPM) levels were considered as equivalent. The above assumptions may add to inaccuracy already inherent in such estimates.

### Carbon Monoxide

Incomplete combustion of fuels produces carbon monoxide (CO). The CO and particle emission pose a serious problem when biomass fuels are used. Smithhas estimated that about 38, 17, 5 and 2 g/meal carbon monoxide is released during the household cooking, using dung, crop residues, wood and kerosene respectively. During the use of liquid petroleum gas (LPG) a negligible amount of CO is released. A studyby the National Institute of Occupational Health (NIOH), Ahmedabad reported indoor air CO levels of 144, 156, 94, 108 and 14 mg/m$^3$ air during cooking by dung, wood, coal, kerosene and LPG respectively. The short-term health effects of CO exposure are dizziness, headache, nausea, feeling of weakness, *etc.* The association between long-term exposure to carbon monoxide from cigarette smoke and heart disease and foetal development has been described by several authors.

### Poly Organic Material and Poly Aromatic Hydrocarbons

Poly organic material is a loose term used to depict a group of chemicals having two or more rings. Of several chemicals included in this group, the PAHs have attracted interest for their possible carcinogenic effects. In addition to PAH, azo and arino compounds have also been found to be potentially carcinogenic. Most other categories of POM are of less environmental interest or are not found in large amounts in organic combustion products.

Polycyclic aromatic hydrocarbons constitute a large class of compounds released during the incomplete combustion or pyrolysis of organic matter. They are often called polynuclear aromatic (PNA) because they contain three or more aromatic rings that share carbon atoms. Benzo(a)pyrene (BaP) is one of the most important carcinogen of the group. Often it is measured to indicate the presence or absence of PAHs although the relationship between BaP content and actual carcinogenicity may be weak. Anthracene and phenanthracene are not carcinogens but methyl additions may render them carcinogenic. PAHs are activated by the hepatic microsomal enzyme system to carcinogenic forms that bind covalently to DNA.

Study by NIOHshowed that the indoor levels of PAH (total) during use of dung, wood, coal, kerosene and LPG were 3.56, 2.01, 0.55, 0.23 and 0.13 μg/m$^3$ of air respectively. These PAH were fluorene, pyrene, chrysene, benzo(a)anthracene, benzo(b)fluoranthene, benzo(k)fluoranthene, benzo(a)pyrene, dibenz(ah) anthracene, benzo(ghi)perylene and indeno(1,2,3-cd) pyrene. All these PAHs except the first three have been classified as possible carcinogens.

**Formaldehyde**

Patel and Raiyanimeasured levels of formaldehyde in indoor environment during cooking by different fuels. The formaldehyde mean levels were 670, 652, 109, 112 and 68 μg/mof air for cattle dung, wood, coal, kerosene and LPG respectively. The formaldehyde is well recognised to be an acute irritant and long-term exposure can cause a reduction in vital capacity and chronic bronchitis. The formaldehyde is well known to form crosslinks with biologic macro-molecules. Inhaled formaldehyde forms DNA and DNA-protein cross-links in the nasal respiratory mucosa. The formaldehyde has been shown to be carcinogenic in a dose dependent fashion in rodents. The studies done in workers occupationally exposed to formaldehyde have consistently (11 of 13 studies reviewed) shown higher incidence of leukaemia. In an epidemiological study in U.K., significantly excess mortality from lung cancer was observed in workers exposed to high levels of formaldehyde.

**Mutagenic activity of the Smoke Particulate Extract**

Microbial tests are widely used as a screening tool for assessing mutagenic potential of chemical substances. The particulate matter in the smoke generated as a result of incomplete combustion of biomass fuels contains a number of organic compounds. To evaluate their carcinogenic potential, it is necessary to screen their mutagenicity through simple and rapid microbial assay as a first step.

Ames assay is simple and sensitive enough t measure mutagenicity of air-borne particulates, so that many researchers have applied this assay to demonstrate the ambient carcinogenic and mutagenic compounds in the extractable organic matter from air-borne particulates. Mutagenic response of complex mixtures of polycyclic organic matter from the combustion of biomass energy fuels was studied using tester strains TA 98 and TA 100 of *Salmonella typhimurium* which can detect the presence of frame-shift and base-pair mutagens. The results indicated that the organic residues of smoke particulates of wood and cattle dung fuels contained direct acting frameshift mutagens and cattle dung contained only direct acting base-pair mutagens while indirect acting frame-shift and base-pair mutagens were found to present in smoke particulates of both the energy fuels.

## DISEASES ASSOCIATED WITH INDOOR AIR POLLUTION

Respiratory illness, cancer, tuberculosis, perinatal outcomes including low birth weight, and eye diseases are the morbidities associated with indoor air pollution.

### Respiratory Illness

The effect of air pollutants in general would depend on the composition of the air that is inhaled which will depend on the type of fuel used and the conditions of combustion, ventilation and duration for which the inhalation occur. The most commonly reported and obvious health effect of indoor air pollutants is the increase in the incidence of respiratory morbidity. Studies by the NIOHon the prevalence of respiratory symptoms in women using traditional fuels (biomass) (n=175) and LPG (n=99), matched for economic status and age, indicated that the relative risk (with 95% C.I.) for cough, and shortness of breath (dyspnoea) was 3.2 (1.6-6.7), and 4.6 (1.2-18.2) respectively.

#### *Childhood acute respiratory infections*

##### *Acute lower respiratory infections*

Acute respiratory infections (ARIs) are the single most important cause of mortality in children aged less than 5 years, accounting for around 3-5 million deaths annually in this age group. Many studies in developing countries have reported on the association between exposure to indoor air pollution and acute lower respiratory infections. The studies on indoor air pollution from household biomass fuel are reasonably consistent and, as a group, show a significant increase in risk for exposed young children compared with those living in households using cleaner fuels or being otherwise less exposed. Some of the studies carried out in India have reported no association between use of biomass fuels and ARI in children. In a case-control study in children under five years of age in south Kerala, where children with severe pneumonia as ascertained by WHO criteria were compared with those having nonsevere ARI attending out patient department, the fuel used for cooking was not a significant risk factor for severe ARI. Non-severe ARI controls may represent the continuum of the cases themselves. Sharma *et al*in a cross-sectional study in 642 infants dwelling in urban slums of Delhi and using wood and kerosene respectively, did not find a significant difference in the prevalence of acute lower respiratory tract infections and the fuel type.

*Upper respiratory tract infections and otitis media*

Studies on the relationship between indoor air pollution and acute upper respiratory infections in children both from developedand developing nationshave not been able to demonstrate the relationship between the two. However, there is strong evidence that exposure to environmental tobacco smoke causes middle ear disease. A recent meta-analysis reported an odds ratio of 1.48 (1.08-2.04) for recurrent otitis media if either parent smoked, and one of 1.38 (1.23-1.55) for middle ear effusion in the same circumstances. A clinic based case-control study of children in rural New York state reported an adjusted odds ratio for otitis media, involving two or more separate episodes, of 1.73 (1.03-2.89) for exposure to woodburning stoves.

***Chronic pulmonary diseases***

*Chronic obstructive pulmonary disease and chronic cor pulmonale*

In developed countries, smoking is responsible for over

80% of cases of chronic bronchitis and for most cases of emphysema and chronic obstructive pulmonary disease. Padmavati and colleaguespointed out to the relationship between exposure to indoor air pollutants and chronic obstructive lung disease leading to chronic cor pulmonale. These studies showed that in India, the incidence of chronic cor pulmonale is similar in men and women despite the fact that 75% of the men and only 10% women are smokers. Further analysis of the cases of chronic cor pulmonale in men and women showed that chronic cor pulmonale was more common in younger women. Chronic cor pulmonale seemed to occur 10-15 years earlier in women. The prevalence of chronic cor pulmonale was lower in the southern states than the northern states of India. This is attributed to higher ambient temperatures during most part of the year allowing for greater ventilation in the houses during cooking. The authors attributed this higher prevalence of chronic cor pulmonale in women to domestic air pollution as a result of the burning of solid biomass fuels leading to chronic bronchitis and emphysema which result in chronic cor pulmonale. Subsequent studies in India confirmed these findings. Numerous studies from other countries, including ones with cross-sectional and case-control designs, have reported on the association between exposure to biomass smoke and chronic bronchitis or chronic obstructive pulmonary disease.

*Pneumoconiosis*

Pneumoconiosis is a disease of industrial workers occupationally exposed to fine mineral dust particles over a long time. The disease is most frequently seen in miners. Cases of respiratory morbidity who did not respond to routine treatment and whose radiological picture resembled pneumoconiosis have been reported in Ladakh.

However, there are no industries or mines in any part of Ladakh and therefore exposure to dust from these sources was ruled out. Two factors considered responsible for the development of this respiratory morbidity were (i) Exposure to dust from dust storms. In the spring dust storms occur in many parts of Ladakh. During these storms the affected villages are covered by a thick blanket of fine dust, and the inhabitants are exposed to a considerable amount of dust for several days. The frequency, duration and severity of these dust storms vary considerably from village to village; (ii) Exposure to soot – due to the severe cold in Ladakh, ventilation in the houses is kept at a minimum. The fire place is used for both cooking and heating purposes. To conserve fuel during non-cooking periods, the wood is not allowed to burn quickly but is kept smouldering to prolong its slow heating effect. The inmates are thus exposed to high concentrations of soot.

The clinico-radiological investigations of 449 randomly selected villagers from three villages having mild, moderate and severe dust storms showed prevalence of pneumoconiosis of 2.0, 20.1 and 45.3% respectively. The chest radiographs of the villagers showed radiological characteristics which were indistinguishable from those found in miners and industrial workers suffering from pneumoconiosis. The dust concentrations in the kitchens without chimneys varied from 3.22 to 11.30 mg/m$^3$ with a mean of 7.50 mg/m$^3$. The free silica content of these dust samples was below 1%. Dust samples sufficient to allow measurement of the dust concentrations could not be collected during the periods of dust storms. A preliminary analysis of the settled dust samples collected immediately after the storms indicated that about 80% of the dust was respirable and the free silica content ranged between 60 and 70%. Detailed statistical analysis of the data showed that the frequency of dust storms, use of chimney in the houses and age were the most important factors related to the development of pneumoconiosis. Thus, the results of medical and radiological investigations positively established the occurrence of pneumoconiosis in epidemic proportion. Exposure to free silica from dust storms and soot from domestic fuel were suggested as the causes of pneumoconiosis. Low oxygen levels or some other factor associated with high altitude may be an important contributory factor in causation of pneumoconiosis because it has been reported that the miners working at high altitude are more prone to develop pneumoconiosis than their counterparts exposed to the same levels of dust and working in the mines at normal altitude.

***Lung cancer***

The link between lung cancer in Chinese women and cooking on an open coal stove has been well established. Smoking is a major risk factor for lung cancer, however, about two-thirds of the lung cancers were reported in nonsmoking women in China, Indiaand Mexico. The presence of previous lung disease, for example tuberculosis which is common in Indian women, is a risk factor for development of lung cancer in non-smokers. The smoke from biomass fuels contain a large number of compounds such as poly aromatic hydrocarbons, formaldehyde, *etc.* known for their mutagenic and carcinogenic activities, but there is a general lack of epidemiological evidence connecting lung cancer with biomass fuel exposure. The factors associated with rural environment may have a modulating effect on the occurrence of lung cancer and therefore the low incidence of lung cancer in Indian women should not lead to a final conclusion of no link between biomass exposure and lung cancer. It may be concluded that at present there is limited evidence of indoor exposure from coal fires leading to lung cancer and there is no evidence for the biomass fuels. Further investigations are needed to reach definite conclusions.

*Pulmonary tuberculosis*

Mishra *et al*recently reported the association between use of biomass fuels and pulmonary tuberculosis on the basis of analysis of data collected on 260,000 Indian adults interviewed during the 1992-93 National Family Health Survey. Persons living in households burning biomass fuels were reported to have odd ratio of 2.58 (1.98-3.37) compared to the persons using cleaner fuel, with an adjustment for confounding factors such as separate kitchen, indoor overcrowding, age, gender, urban or rural residence and caste. The analysis further indicated that, among persons aged 20 years and above, 51% of the prevalence of active tuberculosis was attributed to smoke from cooking fuel.

There is experimental evidence to show that the exposure to wood smoke may increases susceptibility of the lungs to infections. Exposure to smoke interferes with the mucociliary defences of the lungsand decreases several antibacterial properties of lung macrophages, such as adherence to glass, phagocytic rate and the number of bacteria phagocytosed. Chronic exposure to tobacco smoke also decreases cellular immunity, antibody production and local bronchial immunity, and there is increased susceptibility to infection and cancer. Indeed, tobacco smoke has been associated with tuberculosis. Although the evidence in favour of tuberculosis associated with biomass fuel exposure is extremely weak, there is a theoretical possibility of such an association and considering the public health importance of the problem further experimental and epidemiological studies are necessary.

### *Cataract*

During cooking particularly with biomass fuels, air has to be blown into the fire from time to time especially when the fuel is moist and the fire is smouldering. This causes considerable exposure of the eyes to the emanating smoke. In a hospital-based case-control study in Delhi the use of liquefied petroleum gas was associated with an adjusted odds ratio of 0.62 (0.4-0.98) for cortical, nuclear and mixed, but not posterior sub capsular cataracts in comparison with the use of cow dung and wood. An analysis of over 170,000 people in Indiayielded an adjusted odds ratio for reported partial or complete blindness of 1.32 (1.16-1.50) in respect of persons mainly using biomass fuel compared with other fuels after adjusting for socio-economic, housing and geographical variables; there was a lack of information on smoking, nutritional state, and other factors that might have influenced the prevalence of cataract. It is believed that the toxins from biomass fuel smoke are absorbed systematically and accumulate in the lens resulting in its opacity. The growing evidence that environmental tobacco smoke causes cataracts is supportive.

### *Adverse Pregnancy Outcome*

Low birth weight (LBW) is an important public health problem in developing nations attributed mainly to undernutrition in pregnant women. Low birth weight has serious consequences including increased possibility of death during infancy. Exposure to carbon monoxide from tobacco smoke during pregnancy has been associated with LBW. Levels of carbon monoxide in the houses using biomass fuels are high enough to result in carboxyhaemoglobin levels comparable to those in smokers. In rural Guatemala, babies born to women using wood fuel were 63 g lighter than those born to women using gas and electricity, after adjustment for socio-economic and maternal factors. A study carried out in Ahmedabad reported an excess risk of 50% of stillbirth among women using biomass fuels during pregnancy. An association between exposure to ambient air pollution and adverse pregnancy outcome has been widely reported. Considering the association of LBW with a number of disease conditions later in life, there is a need for further studies.

## Intervention Programme in Indoor Air Pollution

Adequate evidence exists to indicate that indoor air pollution in India is responsible for a high degree of morbidity and mortality warranting immediate steps for intervention. The intervention programme should include (i) Public awareness; (ii) Change in pattern of fuel use; (iii) Modification in stove design; (iv) Improvement in the ventilation; and (v) Multisectoral approach.

### Public Awareness

The first and the most important step in the prevention of illnesses resulting from biomass fuels is to educate the public, administrators and politicians to ensure their commitment and promoting awareness of the long-term health effects on the part of users. This may lead to people finding ways of minimising exposure through better kitchen management and infant protection.

### Change in Pattern of Fuel Use

The choice of fuel is mainly a matter of availability, affordability and habit. The gobar gas plant which uses biomass mainly dung has been successfully demonstrated to produce economically viable quantities of cooking gas and manure. Recently, the Government of Andhra Pradesh has introduced a programme called the Deepam Scheme to subsidise the cylinder deposit fee for women from households with incomes below the poverty line to facilitate the switch from biomass to LPG. Such schemes will encourage the rural poor to use cleaner fuels. The use of solar energy for cooking is also recommended.

**Modification in stove design**

Use of cleaner fuels should be the long-term goal for the intervention. Till this goal is achieved, efforts should be made to modify the stoves to make them fuel efficient and provide them with a mechanism (*eg* chimney) to remove pollutants from the indoor environment. Several designs of such stoves have been produced. NIOH study showed significant decrease in levels of SPM, SO, NOand formaldehyde with specially designed smokeless stoves in comparison with traditional cooking stoves. However, they have not been accepted widely. Large scale acceptance of improved stoves would require determined efforts. The most important barriers to new stove introduction are not technical but social.

**Improvement in ventilation**

In many parts of the country poor rural folk are provided with subsidised houses under various government/ international agencies aided schemes. Ventilation in the kitchen should be given due priority in the design of the houses. In existing houses, measures such as putting a window above the cooking stove and providing cross ventilation through the door may help in diluting the pollution load.

**Multisectoral approach**

Effective tackling of indoor air pollution requires collaboration and commitment between agencies responsible for health, energy, environment, housing and rural development.

## Household Energy and Human Health

Poverty remains the greatest barrier to reducing the enormous global inequalities in health that still exist, and which in some countries are worsening. A hitherto neglected factor, which is both a cause and a result of poverty, is the continuing dependence of almost half of the world's population on polluting and inefficient household fuels and appliances. Household energy impacts on the health of the poor through a variety of physical, social and economic routes, but the most important direct health impact results from indoor air pollution (IAP) produced by burning biomass fuels (wood, dung, crop wastes) and coal in simple stoves with inadequate ventilation.

Until quite recently, there has been inadequate official recognition of the size and extent of the problem and little effective action. This is despite a number of key publications, meetings and other activities over the past 15 years. Over the last 30 years or so, work has been carried out in the field of household energy by a wide range of research groups, NGOs, and some international agencies. Much of this activity however was project based, often donor-led, and mainly focused on reducing fuel use, cutting costs to households and protecting the environment from deforestation and erosion. There has been awareness of the connections with health, but in practice this has not been prominent. In addition, very little of this work was co-ordinated, and as a result did not bring about substantial change in policy, donor commitment or - most important - action in those countries and poor communities worst affected.

Recent estimates indicate that around three billion people are still primarily dependent on biomass fuels and coal for their everyday household energy needs, including up to 80-90% in some countries of sub-Saharan

Africa. The overall use of biomass fuel is not falling, and indeed continues to increase among the poorest people who also are turning to cruder forms of biomass, and other waste, to use as fuel.

Evidence on the health effects of indoor air pollution has grown in the last 10 years, and there is now fairly good evidence that exposure increases the risk of acute lower respiratory infections (ALRI) in children, chronic obstructive lung disease (COPD) in adults, and lung cancer where coal is used extensively. In addition, evidence is now emerging of links with a number of other conditions, including tuberculosis, perinatal mortality (stillbirths and deaths in the first week of life), low birth weight, asthma, otitis media, cancer of the upper airway, and cataracts. Although there is now more evidence, it is still deficient in a number of important respects, including imprecise measurement of levels of exposure to pollution, inconsistent definitions of health outcomes, and inadequate attention paid to the influence of other 'confounding' factors. The implication of these limitations is that, although we can be fairly confident that IAP does increase the risk of a number of conditions such as ALRI, we cannot be yet sure by how much. It is estimated that IAP is responsible for nearly two million deaths annually in developing countries and around 4% of the burden of disease. In addition to these direct effects of IAP on health, are other consequences arising from the use of household energy in conditions of poverty. These include burns to children falling onto open fires, ingestion of kerosene stored in soft drinks containers, restrictions on income generating activity, the opportunity costs for women of spending many hours collecting wood fuel as this becomes increasingly scarce, as well as injuries arising from carrying heavy loads and collecting fuel in areas where there is a risk of violence.

A wide range of interventions are capable of contributing to a reduction in indoor air pollution. These include interventions at the source of pollution (improved stoves with cleaner fuels), to the living environment (better ventilation); and to user behaviour (use of pot lids, keeping children away from smoke). For any intervention to be successfully implemented, it should be locally appropriate, and sustainable in the market. Full participation of local people (especially women), appropriate financial support (e.g. micro-credit, avoidance of subsidies), inter-sectoral collaboration at local and national levels, and energy policies that support rather than hinder access by the poor to cleaner fuels, are also important. There is an urgent need to consolidate experience of interventions to date, encourage the evaluation of future interventions in a systematic and comparable way, build capacity in countries, and promote the dissemination of experience in ways that are accessible locally and internationally.

Initial work on assessing the economic costs and benefits of interventions in household energy are encouraging. For example, the cost per DALY saved of an improved stove may be in the range US$ 50-100, comparing favourably with other important health interventions. However, more research is needed to quantify the relationship between exposure and risk of disease, and the health gains resulting from measured reductions in exposure. There is now greater awareness of the inter-relationships between household energy, health and poverty. Poverty is a barrier in the transition to modern fuels, while reliance on such fuels holds back development due to the health burden, and the loss of time and opportunities for economic development.

## Household Energy and Development

### *Household fuels*

Estimates suggest that wood provides around 15 % of energy needs in developing countries, rising to 75 % in tropical Africa. In more than 30 countries, wood still provides more than 70% of energy, and in 13 countries it is over 90 %. Moreover, as the prices of alternatives such as kerosene and bottled gas have increased, the use of biomass fuels has increased. With development there is generally a transition up the so-called 'energy-ladder', towards fuels which are progressively more efficient, cleaner, convenient and expensive. Households typically use a combination of fuels, for example wood for cooking and heating, and kerosene or electricity for lighting, so it is not a simple linear progression. However, households do tend to carry out more tasks with more modern fuels as their socio-economic circumstances improve. The problem is that almost half of the world's population relies predominantly on fuels towards the lower end of this energy ladder, and for many the prospects of moving up it in the short term appear limited.

Households reliant on biomass and coal generally use the fuel indoors, in open fires or poorly functioning stoves, and usually with inadequate venting of smoke. The smoke from biomass fuel and coal contains a large number of pollutants that are dangerous to health, including small particles, carbon monoxide, nitrogen dioxide, sulphur dioxide formaldehyde, and carcinogens such as benzo[a]pyrene and benzene. Studies from Asia, Africa and the Americas have shown that indoor air pollution levels in these homes are extremely high, many times the limits set by the U.S. Environmental Protection Agency (EPA) and WHO. The most important impacts appear to be on acute respiratory infections in children, and chronic lung disease in adults, and in developing countries the overall public health impact of IAP is substantial.

For the majority of poor people, the fire, or hearth, is at the centre of home life. It provides the energy required for cooking, warmth and light, and is very often the focal point of the home with important cultural and spiritual significance. Although smoke is recognised as a nuisance, causing sore, running eyes, coughs and headaches, it can also have utilitarian value. Smoke may be used for curing food, protecting roof timbers from boring insects, waterproofing thatch, and helping to keep biting insects away. Nevertheless, lack of access to more modern fuels and appliances does limit quality of life in a variety of ways:

- Lighting may be restricted and provided only by the fire, candles, or simple kerosene wick lamps which can be a significant source of pollution.
- The lack of light restricts activities in the home, including children's homework, reading and opportunities for income generating activities.
- Lack of access to electricity restricts the use of a wide range of appliances that can contribute to food safety (refrigerators), communication/ education, leisure (radio, TV), and economic activity.

In almost all developing countries it is women who provide fuel for the home and carry out most tasks that require energy use in the home. Studies show that fuel collection takes on average between one half and two hours per day. The time spent collecting wood and other fuels has an 'opportunity cost' for women especially during busy agricultural periods. Carrying large loads of wood exposes women to injury from falls and weight carrying, including fractures and miscarriages, while in areas of war and civil unrest they may also be exposed to violence, and injury from land mines. Their work in the kitchen, often close to the fire, means that of all family members they have the greatest exposure, estimated – for example - at between four and seven hours per day in rural Guatemala.

In general, women also have less control than men over how more modern fuels are used and what appliances are obtained. In South Africa for example, it has been found that once grid electricity is available, men have greater influence in decisions about how the electricity is used in the home and cooking needs may be given lower priority than women would wish.

**Environmental Impact**

*Deforestation and fuel supply.* The use of wood fuel is often thought to be the major cause of deforestation and environmental damage. However, in rural areas wood fuel is usually gathered rather than cut from the trees; and it is agricultural practices, the need for building materials, and population pressure, that are the most important factors in deforestation. Nevertheless, pressure on forests from the use of wood fuel is a problem. Taking South Africa as an example, it is estimated that the three million rural households currently use 10 million tonnes of wood annually, whereas six million tonnes is considered to be sustainably available.

In urban areas, the use of wood requires transport over longer distances, which increases the demand for charcoal. Unfortunately, producing and using charcoal is less efficient than direct use of wood fuel, which leads to forest depletion in rural areas providing fuel to cities. An active approach to forest management can play a part in protecting the supply of fuel and the environment, and help place biomass on a firmer footing as a sustainable, renewable energy source. The use of biomass as a fuel should not necessarily be seen as a negative environmental pressure: what is important is how it is managed.

*Contribution to greenhouse gases.* The low energy use of homes in developing countries means that their contribution to the global output of greenhouse gases is relatively small. In 1995, per capita carbon dioxide ($CO_2$) production was estimated at less than two tonnes for developing countries, compared to 12 tonnes for developed countries and 20 tonnes for the USA. While it is possible to argue that since biomass is renewable, replacement of wood burned should result in a fairly neutral carbon balance (the $CO_2$ released from burning being taken up as new trees grow), the problem is that as stoves used in developing country homes have a low efficiency (around 15%), with (for example) nearly 10% of the energy of wood being lost as products of incomplete combustion (PICs). These PICs include methane, which has a greenhouse effect many times greater than $CO_2$. Modern fuels such as liquefied petroleum gas (LPG) burn with a far higher efficiency, and consequently produce considerably less greenhouse gas emissions. This lower impact of LPG on the global environment, in addition to the greater cleanliness and convenience, should be taken into account in considering the overall balance sheet for various energy options.

### *Smoke pollution*

Studies from countries in Asia, Africa and the Americas have measured levels of indoor air pollution associated with cooking with biomass fuels. Most have measured particles –complex mixtures of chemicals in solid form and droplets. These particles are thought to be the most health-damaging component of smoke pollution (especially the smaller ones which are able to penetrate deep into the lungs). Particles are therefore usually described by size – their effective (or aerodynamic) diameter –which is measured in millionths of a metre (microns).

Particles of up to 10 microns in diameter ($PM_{10}$) have been most commonly measured, although some have looked at all (total) suspended particles (TSP). Recent evidence suggests that the very smallest particles are the most dangerous, therefore some studies have measured particles up to 2.5 microns in diameter ($PM_{2.5}$). Concentrations of particles are expressed as the weight of particles (in micrograms, mg) per cubic metre ($m^3$) of air, thus $mg/m^3$. Only results for $PM_{10}$ will be discussed here.

Typical 24-hour mean levels of $PM_{10}$ in homes using biofuels range from 300 to 3,000+$mg/m^3$. While using an open fire the $PM_{10}$ level can reach 20,000 $mg/m^3$ or more. Yet the EPA standard for average daily (24-hour) $PM_{10}$ is 150 $mg/m^3$ (which should be exceeded only once per 100 days). The annual average should not exceed 50 $mg/m^3$. Most 'western' cities rarely exceed these standards.

In assessing the health implications of indoor air pollution, two measures should be considered:

— The level of *ambient pollution* in the home, which shows how polluted the domestic environment is.

— The level of *exposure* that different members of the family receive, which is a reflection of the time each person spends in polluted environments. A study in rural Kenya, for example found that exposure during brief high-intensity emission 'episodes' (when the fire is stirred or blown upon) accounted for 30 to 60% of the total exposure of household members who took part in cooking and 0 to 11% for those who did not. Simple models that neglect the spatial distribution of pollution within the home, intense emission episodes, and activity patterns may seriously underestimate exposure.

Unfortunately, the measurement of smoke particles in the home is technically quite demanding, involving moderately expensive equipment, careful procedures and quality control and well-trained staff. The measurement of personal exposures adds additional complexity. These methodological complexities are one reason why direct measurement of pollution exposure has rarely been carried out, and why the development of methods for exposure assessment is an important priority for further work in this field.

### Indoor Air Pollution and the Global Burden of Disease

Taken together, estimates of health risk, levels of personal exposure, numbers of people exposed and disease rates, can provide an idea of the overall global burden of disease (GBD) resulting from IAP. This approach is encapsulated in the global burden of disease (GBD) project.

The aim of the GBD project is to help prioritisation in research, policy and resource allocation. Originally the method was applied to disease outcomes, but more recently it has included the disease burden attributable to various risk factors including malnutrition, water and sanitation, tobacco, unsafe sex (risk of HIV/AIDS) and air pollution. For risk factors, the method combines information on level of exposure, level of risk of disease for conditions known to be associated with the risk factor, incidence of disease and the number of people at risk.

The GBD project uses the concept of 'disability adjusted life years' (DALYs) lost. This measure combines deaths and illness, allowing for the level of disability (impaired health) resulting from the illness and the number of years of life affected by this disability (if the person survives) or lost completely (if the person dies). A condition such as ALRI which has a high mortality in very young children would contribute significantly to the DALYs measure, since many years of potentially healthy life are lost if a young child dies.

There are currently four methods for estimating the burden of disease from the use of solid fuels in developing countries: the pollutant based method, child survival curves, cross-national (ecological) comparisons and the exposure-based method. The results from each method are fairly similar, which provides some credibility for the approaches taken and confirms the extent of the public health problem of IAP in developing countries.

The table below present results derived from the exposure-based method, as this would appear to be the most valid and generally applicable for developing country populations.

*Table 1: Burden of disease from solid fuel use in the early 1990s using the exposure based method*

| *Region* | *Deaths* | *Illness Incidence* | *DALYs* |
|---|---|---|---|
| India | 496,059 | 448,351,369 | 15,954,430 |
| China | 516,475 | 209,727,474 | 9,335,387 |
| Other Asia & Pacific Islands | 210,721 | 306,356,582 | 6,599,471 |
| Sub-Saharan Africa | 429,027 | 350,703,204 | 14,323,188 |
| Latin America | 29,020 | 58,246,497 | 918,236 |
| Middle East and North Africa | 165,761 | 64,150,732 | 5,633,022 |
| LDC Total | 1,800,000 | 1,400,000,000 | 53,000,000 |

*Table 2: Percentage of total LDC burden attributable to solid fuel use*

| *Region* | *Deaths* | *Percent ARI* | *DALYs* | *Percent ARI* |
|---|---|---|---|---|
| India | 5.3% | 81 | 5.5% | 87 |
| China | 5.8% | 25 | 4.5% | 50 |

| | | | | |
|---|---|---|---|---|
| Other Asia & Pacific Islands | 3.8% | 75 | 3.7% | 85 |
| Sub-Saharan Africa | 5.2% | 85 | 4.9% | 90 |
| Latin America | 1.0% | 71 | 0.9% | 82 |
| Middle East and North Africa | 3.6% | 89 | 3.7% | 93 |
| LDC Total | 4.7% | 67 | 4.3% | 81 |

**Economic Evaluation of Household Energy Interventions**

For any intervention to be successful, those for whom it is intended need to believe it is worth making a change. Households can be expected to adopt a new technology if the perceived benefits of adoption are greater than the costs. While the physical impacts of adopting an improved stove, such as reduced emissions or improved fuel efficiency, can be observed directly, the value in monetary terms to the household is less evident. Estimating monetary value is useful if we are to understand the household benefits of potential interventions and, therefore, household incentives for adoption. A clear understanding of such incentives helps identify supporting policies that increase—and conflicting policies that reduce—household benefits of implementing the interventions; and provides important evidence for decisionmakers weighing up the 'value for money' offered by various interventions for health improvement.

Research indicates that interventions in household energy are potentially attractive from an economic standpoint. We know, for example, that:

— Cost-benefits studies suggest a favourable cost to benefit ratio for existing interventions.

— Cost-effectiveness estimates carried out to date estimate a cost of $50-100 per DALY saved, based on data from India. This compares well with a range of other preventive and curative health interventions.

Households make production, consumption, and time-allocation decisions to maximise their welfare, which depends on (amongst other factors) child and adult health. Their decision-making is affected by:

- Constraints that are external to the household, such as market prices.
- Household technological constraints on crop production, fuel production, the production of health, etc.
- The household budget constraint, because cash expenditures must equal cash income.
- Constraints imposed by household endowments, including land, labour time, and so on.
- The household's time constraint 'allows' available female labour time to be allocated to fuel collection, crop production, cooking, and other household-related activities.
- Constraints imposed by information the household possesses about technologies, health impacts, etc.

The costs of adopting an intervention are usually clear, but the benefits are not. How then can benefits be identified and measured?

Benefits are all changes in household welfare due to the outcome of the intervention and costs are the value of the inputs required to adopt and use the intervention. Benefits of interventions to control indoor air pollution are here assumed to include direct improvements in adult and child health and indirect improvement in child health generated by better adult health.

**Costs and benefits**

The costs and benefits have been estimated for mortality and morbidity (ALRI) using examples of interventions from Guatemala and a ceramic chimney-less stove from

Kenya. Briefly, a three-step method has been used to compute the cost-benefits of interventions to control indoor air pollution:

1. Estimate the three types of health impacts (direct child and adult and indirect child).
2. Estimate the monetary value to the household of these health impacts.
3. Compare the monetary value to the costs of the intervention.

These examples indicate that for mortality (all cause) the benefits exceed the stove costs by a factor of at least five. For morbidity (ALRI incidence) the findings indicate that if there are several children in a household (which is typically the case) the direct household benefits of reduced ALRI in children alone could justify an investment in an improved stove. Why, then, have so few households adopted the interventions? Explanations include:

— Practical problems with the improved stoves.

— Additional costs. For example, loss of the social benefits of a traditional stove.

— Lack of information. Households may not know or may not understand the relationship between fuel use, cooking technology, and household health.

— Poverty. Household valuation of risk reduction in developing countries is constrained by income levels that are substantially lower than those found in developing countries.

### Towards Effective and Sustainable Interventions

There has been little systematic evaluation of the direct impacts (e.g. reduction of IAP exposure, safety) and indirect impacts (e.g. opportunity costs of women's time, environmental impacts, etc.) of these potential interventions on health. Available information relates mainly to the impact on fuel consumption, emission levels, and in fewer cases on the resultant pollution and exposure levels. Therefore current knowledge is almost entirely restricted to source interventions, and mainly for various types of improved stove. Experience regarding the acceptance and sustainability of household energy programmes has increased, although these aspects have rarely been the subject of systematic research.

The emphasis in the 1970s was on fuel efficiency, which brought with it a somewhat narrow focus on technological solutions such as improved biomass stoves. These stoves included enclosed mud devices, often with chimneys, which were largely unsuccessful because of their low efficiency and rapid deterioration. A large-scale stove programme in India appears to be suffering from these problems, although a thorough evaluation is not available. Furthermore, work on the benefits of improved stoves was often marked by a lack of data on stove performance in everyday use. Efficiencies and emissions, for example, were commonly measured in controlled environments by technical experts under conditions very dissimilar to those in the field. More recently there have been evaluations of efficiency and pollution levels in more typical everyday circumstances in the home environment.

Recent surveys have identified several hundred improved stoves programmes in over 50 nations. They range from entirely local, non-governmental advocacy to national initiatives reaching millions of households. The quality and efficiency of individual stove designs has also varied greatly, as has the success of the programmes. This lack of success has shifted the focus to identifying factors that can result in successful technical design, and dissemination of improved stoves. For example, a more successful approach has been with ceramic chimney-less stoves which are cheap, relatively durable, and more fuel-efficient. Such stoves have been quite popular, especially where sustainable markets have developed. They can reduce indoor air pollution because of better combustion, with lower emissions and potentially also shorter cooking time, although the reduction in IAP is generally no more than 50% at best.

Levels of indoor air pollution, exposure, the subsequent health impact and potential solutions are all highly dependent on the local context and the specific needs of a particular household. If interventions are to be successful they must be sensitive to local conditions and build on the particular ways in which the people exposed to high levels of pollution in the home respond to the problems they face, and the opportunities they have for change. The key to success is to broaden the range of secure and sustainable choices available to the local actors in devising solutions.

### Characteristics of Successful Intervention

Tools are needed to ensure the characteristics of successful interventions outlined above are taken into account in local planning, and in the way that interventions and policies are evaluated. As a basis for such a tool, nine criteria have been proposed to reflect the broad range of health, environmental and development issues identified as being relevant to household energy use in poor countries and the overriding goal of improving health. Fuel efficiency *per se* is not included, as it is thought to .be more usefully considered through its contribution to operating costs, exposure and environment.

These criteria can be applied in a similar way to a health impact assessment, although in this situation the focus is primarily on household energy. The criteria are intended to be relevant both to *rapid local option appraisal,* as well as more *systematic evaluation at national and international levels.* At a local level, criteria may be used to help inform choices through a broad-based assessment of likely impacts on health. The criteria should not be applied too rigidly, as priorities will vary and information will often be unavailable. However, discussion of the criteria should ensure that all key issues, including those impacting directly on health, are at least considered by all concerned. At an international level, the criteria would be useful for formal comparative evaluation, as part of national and international efforts to compile experience with interventions using relatively well-standardised criteria and methods. Information from existing studies is quite limited, but this would be built up as further evaluation is carried out using agreed criteria and methods and would provide the basis for a reference resource. Findings from comparative evaluations so far undertaken have shown both the potential local usefulness of the method, as well as the need for integration of local assessment with international collaborative assessment of health impacts.

Biogas is very beneficial from an exposure level, cost, and environmental point of view, but is limited to a specific set of circumstances – namely where cattle are kept, dung is easy to collect, and water is available. Alternative fuels, such as ethanol and biodiesel, may be produced from energy crops such as sugar cane and rape seed and are thus non-fossil fuels, but as liquids benefit from easier combustion conditions for complete combustion. Ethanol may be socially unacceptable in some areas because of alcohol taboos, but this may be overcome by the production of *gel fuels,* which are currently being developed and evaluated.

*Interventions to the living environment.* From data that is available, the benefit of a cooking window (a type of enclosed 'fume cupboard') as based on findings of

Nyström was notable. Work in rural Kenya is assessing the usefulness of hoods (with flues), enlarging and repositioning windows, and enlarging eaves spaces. Developed with full participation of local women, these interventions are acceptable and quite robust. They are also quite effective at reducing pollution and personal exposures of women, especially the use of hoods that are able to vent smoke to the exterior of the house. Further work to develop and assess the market readiness of these interventions among the poor rural communities concerned is currently underway.

*Interventions to user behaviours.* Objective information about user-based interventions such as drying wood, or protecting the child from smoke, is scarce, although proposals for further research are being developed.

However, reductions through changes in the behaviour are unlikely to bring about exposure reductions as large as can be expected from a fuel switch or installation of a hood, chimney or cooking window. So user-based interventions should perhaps be seen as important supporting measures for interventions in other subsystems, but further work is needed.

# 3

# Outdoor Air Pollution

Air pollutants were among the first environmental contaminants to be identified and monitored. Air pollutants can be categorised as primary or secondary pollutants depending on their origin. Primary pollutants include gases such as sulphur dioxide ($SO_2$) and nitrogen oxides ($NO_x$), and particulates such as lead and carbon. Secondary pollutants are formed in the air when various gases and particulates interact with each other in the presence of sunlight. These include gases such as ground-level ozone ($O_3$), commonly known as smog, and sulphuric acid ($H_2SO_4$), which is commonly known as acid rain. Ground-level ozone, sulphur dioxide, nitrogen dioxide, carbon monoxide and particulate matter have been of greatest interest to Health Canada researchers because of their direct or indirect effects on the cardio-respiratory system.

Ground-level ozone ($O_3$) is a gas found in the low atmosphere that is formed when two kinds of pollutants, nitrogen oxides and volatile organic compounds (VOCs), react in the presence of sunlight. High concentrations can be found during hot, still summer weather. Ground-level ozone is the primary component of "photochemical smog" and is different from the blanket of ozone high above the Earth, which protects us from the sun's harmful ultraviolet (UV) rays. When levels are present above normal background levels, it can cause adverse cardio-respiratory health effects, as well as damage to vegetation and other materials.

Sulphur dioxide ($SO_2$) is a gaseous pollutant that can be chemically transformed in the atmosphere in the presence of other chemicals and sunlight to form acidic pollutants such as sulphuric acid and sulphates.

Nitrogen dioxide ($NO_2$) is one of a group of gases called nitrogen oxides, which are composed of nitrogen and oxygen. Like sulphur dioxide, nitrogen oxides can react with other chemicals in the atmosphere in the presence of sunlight to form acidic pollutants, including nitric acid.

Particulate matter ($PM_{10}$, $PM_{2.5}/_{2.1}$) refers to small solid or liquid particles suspended in the air that can be formed as a result of natural, industrial or other human activities. Particulate matter is measured as TSP, which includes all sizes of particles, $PM_{10}$ and $PM_{2.5}/_{2.1}$. The finer particles are of most concern because these are small enough to travel deeply into the lungs. The particles themselves cause health concerns, but they can also carry chemicals into the lung, e.g., polycyclic aromatic hydrocarbons (PAHs), metals, etc.

Volatile organic compounds (VOCs) are compounds containing carbon that evaporate quickly at ordinary temperatures. Volatile organic compounds react with nitrogen oxides to form ground-level ozone.

Carbon monoxide (CO) is a colourless, odourless, poisonous gas caused mainly by incomplete combustion. It is emitted mainly from vehicles.

## Sources of Outdoor Air Pollution

The primary source of ambient (outdoor) air pollution is

the burning of fossil fuels such as coal, oil, gasoline and diesel fuel for electricity, heating and transportation. Vehicle exhaust from private and commercial vehicles is a major contributor of nitrogen oxides, carbon monoxide, VOCs, and particulate pollution in urban areas. Pollutants (e.g., sulphur dioxide and nitrogen oxides, particulate matter) are also released by industries such as pulp and paper mills, ore smelters, petroleum refineries, power generating stations and incinerators. Other activities that generate pollutants include home heating and the use of appliances requiring oil, electricity, wood, natural gas or other fuels. Fumes and particulate matter can also be released into the air from naturally occurring events such as volcanoes, wind-driven soil erosion and forest fires.

Air pollution does not originate only from local sources. Pollutants can be carried thousands of kilometres from their original source, sometimes from urban areas to less populated areas. This process is generally referred to as "transboundary pollution" or "long-range atmospheric transport." For example, residents of southwestern Ontario are exposed to pollutants carried by wind patterns from large metropolitan centres, such as Detroit, Michigan. Pollutants that originate in the industrial Ohio River Valley of the United States also frequently end up in the Great Lakes basin, particularly in the Windsor–Quebec corridor, where they contribute to ground-level ozone and particulate air pollution. Even areas remote from urban and industrial centres, such as the Arctic, can be affected by air pollutants carried there by weather patterns.

Additional air pollutants of concern to human health include polychlorinated biphenyls, dioxins and other persistent organic compounds that are transported long distances by air and are deposited in bodies of water and on land. Their persistence allows them to bioaccumulate in the food chain where they can then pose a threat to people consuming large amounts of contaminated selfcaught fish and wildlife.

Other "air toxics" of concern comprise a broad range of substances including toxic metals (i.e., lead, mercury, cadmium, arsenic, chromium and nickel) and compounds such as benzene, formaldehyde and trichloroethylene. These pollutants are monitored through Environment Canada's National Air Pollution Surveillance (NAPS) program.

### Health Effects of Air Pollution

There is strong evidence from research on people living in the Great Lakes basin that increases in certain air pollutants are linked to increases in hospitalisation for respiratory and cardiac disease, and to increased premature mortality. Most importantly, there does not appear to be a "threshold level" for ground-level ozone or particulates below which no effects are observed; even low levels of ground-level ozone and particulates are damaging to the cardio-respiratory system.

Individual reactions to air pollutants depend on the type of pollutant, how much of the pollutant is present, the degree of exposure and the kinds and levels of individual activity (people working or exercising outdoors have greater exposure). Age and health status are also very important. In general, the more susceptible populations include the elderly and those with cardiac or respiratory diseases such as asthma, emphysema and chronic bronchitis. Children also tend to be more sensitive than adults because they breathe faster, and in summer spend more time being active outdoors.

Health effects may range from difficulty in breathing, through coughing and wheezing, to aggravation of existing cardiac and respiratory conditions. These effects may result in visits to a doctor or an emergency room, increase use of medications, admission to hospital or even premature death. For sensitive individuals, any increase in air pollution, no matter how small, could cause their illness to become more severe.

**Respiratory and Cardiac illness**

There is strong evidence from the Great Lakes basin that link ozone, airborne particles and acid aerosols to significant respiratory health effects including illness requiring hospital admission and even premature mortality. There is also evidence from the Great Lakes basin that these pollutants cause reduced lung function in children. Although results from the study of hospital admissions for cardiac disease suggest an increase in disease with outdoor sulphate levels, it is unclear exactly how the effect occurs.

## Reducing Outdoor Air Pollution

Action has been taken on a wide range of air pollution problems. Some highlights:

— Improvements in vehicle emissions standards have helped reduce the amounts of air pollutants released by individual vehicles. Vehicle emissions are controlled by the federal *Motor Vehicle Safety Act*, which sets limits on tailpipe emissions of carbon monoxide (CO), nitrogen oxides, hydrocarbons (HCs) and diesel particulates for new vehicles. Mandatory vehicle emissions testing currently being considered in Ontario would significantly reduce air pollution. While car pooling and the use of public transportation reduces the levels of pollutants emitted into the air, the number of cars and trucks in use in Canada continues to increase, and thus the total amounts of most air pollutants have decreased only gradually.

- In 1991, Canada and the United States signed the Canada-U.S. Air Quality Agreement, in which the two countries agree to work together to address the issue of trans-boundary air pollution. The Canadian government works closely with industry on both sides of the border to encourage voluntary reductions and more stringent environmental controls on the types and amounts of pollutants that can be emitted into the atmosphere. Reductions in emissions of several air pollutants have been observed in parts of the Great Lakes basin because of the efforts of some industrial sectors.
- The $NO_x$/VOCs Management Plan was established in 1989 in order to target the reduction of ground-level ozone. Nitrogen oxides ($NO_x$) and volatile organic compounds (VOCs) are the primary precursors of ground-level ozone.

As a result of Canada' s ozone protection programs and Canada' s obligations under the *Montreal Protocol on Substances that Deplete the Ozone Layer,* as of January 1, 1996, Canada has prohibited the production and import of newly produced CFCs and other ozone depleting substances (ODS) except for certain essential uses such as CFCs for metered dose inhalers for the treatment of asthma and chronic obstructive pulmonary disease. Environment Canada regulates the production, use, import and export of ODS and certain products containing ODS under the *Canadian Environmental Protection Act* (CEPA).

The provinces are responsible for implementing programs for recovery and recycling of ozone-depleting substances. Nine provinces have regulations on the recovery and recycling of ODS and the Yukon and

Newfoundland are currently developing theirs. Environment Canada chairs a federal-provincial working group that provides support for implementation of the recovery, recycling and reclamation of CFCs.

The Program on Energy Research and Development (PERD) was established in 1974 to ensure that knowledge and technology options are available for Canada to develop and use energy resources in a cost-effective and environmentally responsible manner. Health Canada's involvement in the program began in 1981.

## Air Monitoring and Regulation

### Air Quality Objectives (AQOs)

An air quality objective is a statement of the concentration of an air pollutant and the period of exposure to it, which, if not exceeded, affords a specified degree of protection to humans, animals, plants and materials. Objectives are generally set to cover both short- and long-term exposure to air pollutants.

To date, Canada has had a three-tiered system of National Ambient Air Quality Objectives (NAAQOs). NAAQOs were established under this system for sulphur dioxide ($SO_2$), carbon monoxide (CO), nitrogen dioxide ($NO_2$), total suspended particulates (TSP), and ground-level ozone ($O_3$). These pollutants have been measured daily by a network of air monitoring stations throughout Ontario. These monitoring stations are operated by both federal and provincial agencies. Average or maximum concentrations may be reported over a number of time periods, depending on the relevant provincial objectives or criteria for the pollutant.

The three-tiered system of NAAQOs is being replaced by a more science-based two-tiered system. In this system, the Reference Level is a science-based level, above which there are demonstrated effects on human health and/or the environment. The Air Quality Objective is a level that is based on a combination of scientific, social, technological, and economic information, and represents the air quality management goal for the protection of the general public and the environment in Canada. New NAAQOs are currently being developed under this system, beginning with particulate matter and ozone, which are expected in 1998.

The NAAQOs are established based on recommendations from a Federal-Provincial Advisory Committee and Working Group on Air Quality Objectives and Guidelines. Provincial governments have the option of adopting these either as objectives or as enforceable standards using their legislation. The provinces and the federal government have recently agreed to develop the NAAQOs into Canada-wide standards.

*Table 1: National Ambient Air Quality Objectives for Selected Pollutants (Maximum Acceptable Concentration)*

| *Pollutant (units)* | *Period of Exposure* | | | |
|---|---|---|---|---|
| | *1-hour* | *8-hour* | *24-hour* | *annual* |
| Sulphur dioxide (ppb) | 340 | - | 110 | 20 |
| Ozone (ppb) | 82 | - | - | 15 |
| Suspended particulate matter (μg/m3) (TSP) | - | - | 120 | 70 |
| Carbon monoxide (ppm) | 31 | 13 | - | - |
| Nitrogen dioxide (ppb) | 210 | - | - | 50 |

Air quality is monitored by both provincial governments and by the federal government. Federal involvement is to ensure a nationally coordinated monitoring system and to meet international agreements. Provincial governments monitor primarily for compliance with emission control guidelines. The federal government has also established regulations for the release of vinyl chloride, asbestos, lead and mercury from specific stationary sources such as secondary smelters or vinyl chloride plants. Vehicle emissions are regulated by the federal *Motor Vehicle Safety Act*, which sets limits on tailpipe emissions of carbon monoxide (CO), nitrogen oxides (NOx), hydrocarbons (HCs) and diesel particulates for new vehicles only.

In Ontario, several regulations govern ambient air quality. Regulation 309 defines air quality criteria for 15 pollutants that form the basis of the Ontario Air Pollution Index; Regulation 296 defines Ambient Air Quality Criteria (AAQC) for 23 pollutants; and Regulation 312 limits the amount of sulphur in fuels.

# 4

# Urban Air Pollution

Quantifying the magnitude of the impact of air pollution in cities worldwide, however, presents considerable challenges owing to the limited availability of information on both effects on health and on exposures to air pollution in many parts of the world. Measurements of urban air pollution, when available, are available largely for a nonrepresentative sample of urban areas. Many areas of the world lack measurements of any kind, and these must then be estimated using statistical models. On the basis of these considerations, we defined the target population for this risk assessment exercise as the residents in the year 2000 of national capital cities and of cities worldwide with populations of >100000.

Man-made urban air pollution, which is derived largely from combustion processes, is a complex mixture containing many toxic components. We indexed this mixture in terms of PM, a component that has been consistently linked with serious effects on health, and, importantly, the levels of which can be estimated worldwide. Exposure to PM has been associated with a wide range of effects on health, but its effects on mortality are arguably the most important, and are also most amenable to global assessment.

## Urban Air Pollution from Combustion Sources

Combustion of fossil fuels for transportation, power generation, and other human activities produces a complex mixture of pollutants comprising literally thousands of chemical constituents. Exposure to such mixtures is a ubiquitous feature of urban life. The precise characteristics of the mixture in a given locale depend on the relative contributions of the different sources of pollution, such as vehicular traffic and power generation, and on the effects of the local geoclimatic factors. The relative contribution of different combustion sources is a function of economic, social and technological factors, but all mixtures contain certain primary gaseous pollutants, such as sulfur dioxide ($SO_2$), nitrogen oxides ($NO_X$) and carbon monoxide (CO), that are emitted directly from combustion sources, as well as secondary pollutants, such as ozone ($O_3$), that are formed in the atmosphere from directly-emitted pollutants. The pollutant mixture also contains carcinogens such as benzo(a)pyrene, benzene and 1,3-butadiene. When petrol contains lead (Pb), as is still the case in many developing countries, this element is a common constituent of the pollution mix.

All combustion processes produce particles, most of which are small enough to be inhaled into the lung either as primary emissions (such as diesel soot), or as secondary particles via atmospheric transformation (such as sulfate particles formed from the burning of fuel containing sulfur). Their concentrations (in micrograms per cubic metre, or $\mu g/m^3$ ) are generally measured as inhalable and fine particles, $PM_{10}$ and $PM_{2.5}$, respectively. However, the total suspended particle mass (TSP) is still the only particle measurement available in many developing countries.

Pollution from the combustion of fossil fuels is largely emitted into the outdoor air, but human exposure occurs both indoors and outdoors. An individual's exposure to ambient urban air pollution depends on the relative amounts of time spent indoors and outdoors, the proximity to sources of ambient air pollution, and on the indoor concentration of outdoor pollutants. The indoor concentrations depend on factors such as the circulation of the indoor air and the degree to which constituents of the outdoor combustion mixture penetrate and persist in the indoor environment. Studies conducted largely in Europe and North America have shown that the fine particles generated from combustion outdoors both effectively penetrate and persist in many indoor environments. Gases, such as sulfur dioxide and ozone, may penetrate the indoor environment, but generally do not persist because of their reactivity. In some rural areas of developing countries, indoor cooking on unvented coal- or biomass-burning stoves is the most significant exposure to pollution from combustion sources. The burden of disease caused by such exposure is addressed in chapter 18. The actual dose delivered to the lung or other organs will further depend on the type of pollutant, the breathing pattern and physical characteristics of the individual that determine the extent and site of deposition.

Governments in many parts of the world monitor ambient concentrations of air pollution as part of regulatory programmes designed to protect public health and the environment. The most extensive monitoring systems are in the United States and western Europe, where regular monitoring of ambient air quality has been in place since the mid-1970s. The most frequently and routinely monitored air pollutants include sulfur dioxide ($SO_2$), nitrogen oxides ($NO_x$, including NO and $NO_2$), carbon monoxide (CO), ozone ($O_3$), lead (Pb), black smoke (BS) or soot, and PM. National monitoring systems also exist in other parts of the world, but access to the data collected by these systems and international standardisation of the monitoring methods are limited. The World Health Organisation (WHO) Air Management Information System (AMIS) collects the available information, but the reporting from many regions is poor, and for some regions there are no data in the WHO database. The various designs of the networks, differences in monitoring objectives and limited availability of the collected data for the outside users limit access to the information on population exposure in the greater proportion of the world's cities. In some parts of the world (e.g. in most of the countries of the former Soviet Union), the monitoring systems exist but do not provide the data necessary for assessment of the impact on health.

These monitoring systems currently provide much of the data on exposure to urban air pollution that have been used in epidemiological research, although some studies establish their own monitoring networks when routinely-collected data are either unavailable or of poor quality, or to measure specific air pollution constituents, such as specific known carcinogens. Typically, monitoring sites are located in the city centre or throughout a given metropolitan area, in order to more accurately reflect the average residential exposure of the population. The data from monitors sited so as to measure emissions from specific sources, such as a local industry or heavy vehicular traffic, are frequently excluded from the data sets, as they may significantly deviate from the average levels of exposure experienced by the population.

Exposure estimates that rely exclusively on data from one or more stationary monitoring sites may provide inaccurate estimates of the short- and/or long-term average personal exposures of study populations. The direction and magnitude of the errors that will be induced in estimates of the relative risk attributable to exposure to air pollution depend on the precision of the air quality monitoring data (or models used to generate the estimates of the concentration of pollution), the applicability of one estimate to the entire target population and the correlation of the errors with the health outcome. Generally, such errors will be smaller for pollutants that tend to be uniformly distributed over large urban areas, and that penetrate efficiently indoors, both of these features being the case for fine PM produced by combustion. If the errors in the estimates of exposure are uncorrelated with the risk of the health outcome, then the estimates of relative risk attributable to air pollution will, in most cases, be too low (i.e. biased to the null).

**Air Pollution Metric for Exposure Variable**

We selected $PM_{10}$ and $PM_{2.5}$ as the indicators of exposure to urban air pollution from combustion sources. As noted above, PM is a ubiquitous component of the mixtures emitted into, and formed in, the ambient environment by combustion processes, and indicates the presence of these mixtures in outdoor air. Most importantly, these measures of particulate air pollution have been used in many epidemiological studies from around the world, of both mortality and morbidity of air pollution, and so provide the best overall indicator of exposure for our purposes. Although other components of ambient air pollution from combustion sources are associated with these and other effects on health, particulate air pollution has been found to be consistently and independently related to the most serious effects of air pollution, including daily and longer-term average mortality. There is some evidence, although much less than that for PM, linking ozone to premature mortality, particularly during the summer months. However, despite recent progress in developing models to estimate tropospheric (ground-level) ozone on a global scale, it was not currently feasible to derive the subregional estimates that would have been required for this project. In many developing countries, exposure to lead in the ambient air may also be of great consequence, having effects on mortality perhaps via effects on blood pressure.

PM has been linked to serious effects on health after both short-term exposure (days to weeks), and more prolonged exposure (years), although there remains some uncertainty as to the distribution of induction times with regard to mortality. We chose the annual average concentration(s) of PM as the exposure metric(s) because it corresponds to the time-scales of *a priori* interest for estimates of attributable and avoidable burden in the Global Burden of Disease (GBD) project, and because it was used to estimate the effects of exposure to PM in the key epidemiological study that provides our estimates of the concentration–response function.

### *Air Pollution Measurements used in Estimating Annual Average Concentrations*

The availability of measurements of ambient concentrations of PM varies widely across the globe, making estimation of annual average concentrations a considerable challenge. To estimate ambient PM concentrations for all 14 subregions, we used a model (Global Model of Ambient Particulates [GMAPS]) recently developed at the World Bank to estimate concentrations of $PM_{10}$ in cities, on the basis of available measurements of PM at population-oriented monitoring sites. The model incorporates information on factors such as fuel mix, level of economic development, demographics and weather, in order to predict ambient concentrations of $PM_{10}$ in urban residential areas. These estimates of $PM_{10}$ were converted to $PM_{2.5}$ using available information on geographic variation in the ratio of $PM_{2.5}$

to $PM_{10}$. For each PM metric, the population-weighted subregional annual average was derived using the population of each city within each subregion in the year 2000.

The GMAPS model developed at the World Bank can be used to generate estimates of concentrations of $PM_{10}$ in all world cities with populations of >100000, and in national capitals. The estimation model is based on available measurements of $PM_{10}$ and TSP from population-oriented monitoring stations in cities worldwide for the period 1985 to 1999, retrieved in October 2001. In all cases, data from a monitoring site were included if and only if it was clearly identified as a residential or mixed residential site. For instance, city averages reported for many Chinese cities were not included in the model estimation because the location of these sites could not be ascertained.

In principle, the monitoring data used for calculation of annual averages should be collected throughout the year, since seasonal patterns in the data are fairly common. More than 85% of cities in Europe and the United States collect measurements of PM throughout the year. The representativeness of the data for cities in other parts of the world could not be confirmed. In addition, in many countries where PM was measured throughout the year, it was only measured on every sixth day. The methods for measuring concentrations of PM also varied, both gravimetric and automatic methods being included. Most of the data on annual average ambient concentrations used in the model come from AMIS. This information is submitted to WHO by national environmental agencies and air quality control authorities, which perform these measurements using nationally approved methods and standards of data quality. The data set contains the annual mean concentration of selected air pollutants, including PM, by monitoring site. Additional data, such as 95$^{th}$ percentiles of daily means, are also available for some sites. Although WHO requests that all Member States provide data for compilation in the AMIS database, the reported data are still limited because many countries do not have air quality monitoring networks. Additionally, some countries with monitoring networks may not report the data because of poor data quality or limited ability to process and report the data.

The data from AMIS were supplemented with other sources of data on TSP and $PM_{10}$ from monitoring sites. These included data for European cities collected by WHO/European Centre for Environment and Health (ECEH) for the Health Impact Assessment of Air Pollution (HIAAP) project in 1999 from both national and local environmental agencies, data for Canadian cities provided by Environment Canada and statistics Canada, and data for cities in the United States from the U.S. Environmental Protection Agency AIRS database. Data for Chinese cities were also obtained from the Environmental Quality Reports from China, and Mexican cities from the Instituto Nacional de Ecología (INE), SEMARNAP, Mexico. Additional data were also obtained from the World Bank URBAIR studies of air pollution in Jakarta and Kathmandu. To limit undue influence of the data from cities in the United States, data used from the United States AIRS database were limited to the years 1996–1999.Measured annual average concentrations of $PM_{10}$ and TSP data from monitoring sites were available for 512 unique locations in 304 cities in 55 countries over the period 1985–1999, and provided 1997 time–location data points. For some sites and years, data on both TSP and $PM_{10}$ were available, yielding a total of 2344 individual observations. A total of 304 cities reported either the annual average concentrations of $PM_{10}$ or TSP for at least 1 year between 1985 and 1999. Of these, 51 cities reported both $PM_{10}$ and TSP while 165 cities, mostly in North America and western Europe, reported $PM_{10}$ only, and the remaining 88 cities reported data for TSP only.

Coverage of cities and populations with data from monitoring sites varies significantly across different subregions. For instance, data from monitoring were available for fewer than two cities for six of the subregions, AFR-D, AFR-E, AMR-D, EMR-B, EMR-D and SEAR-B. In contrast, data from monitoring sites were available for 218 cities in NAWE, of which 174 report data on $PM_{10}$. The 304 world cities with data from monitoring account for 9% of the total number of cities with a population of >100000 worldwide and have a combined population in the year 2000 of around 559 million, or about 28% of the global urban population.

***Estiamting Ambient Concentrations of particulate matter in cities***

The average subregional ambient concentrations used in this work are estimated from the city-level model predictions for 1999, the latest year for which all of the explanatory variables were available. The estimates of concentrations of $PM_{10}$ in each city for 1999 were generated using a three-step approach.

Finally, for cities with actual data from monitoring, a best estimate of concentration of $PM_{10}$ in the city in 1999 was generated by incorporating information on concentrations from previous years. Specifically, an average residual for each city was determined by comparing each yearspecific predicted value generated by the estimation model (1) with the actual monitored value for that year and city. This served to adjust the model predictions for local factors that are known but unmeasured in the model, such as the composition of local economic activity. Given the large year-to-year variations in the available measured data even at the same monitoring station, correcting for the average residual provides a better representation of long-term average factors affecting concentrations of PM in a city than using the actual monitored value for the last year of data from monitoring alone.

***Estimating Subregional ambient concentrations of particulate matter***

To avoid extrapolating outside the sample frame, all exogenous variables were truncated to the range used in the estimation sample. When necessary, missing explanatory variables for the country were filled in with the median values for economically similar countries located in the same geographic area. For most subregions, data were available for at least 95% of the cities, accounting for at least 95% of the population in each subregion. In contrast, data on either fuel, GDP or gross national product (GNP) were missing for 20–30% of the cities, accounting for 20–30% of the population for each of the four subregions AFR-D, AFR-E, EMRB and EMR-D. In all, data on either fuel, GDP or GNP were completed in this way for 176 cities worldwide, accounting for 5% of the total world urban population.

The estimated annual average concentrations of $PM_{10}$ in urban areas for world cities with populations of >100000. Standards currently in place in North America and western Europe lie between 30–60μg/m$^3$ . Therefore, we defined a middle group with concentrations of $PM_{10}$ in the range of 30–60μg/m$^3$ . Cities with values that fell outside this range were sorted into two groups of cities with higher concentrations and two groups with lower concentrations. Worldwide, about 30% of the urban population live in the less polluted cities while 40% live in the more polluted cities. The remaining 30% of people live in cities with concentrations of $PM_{10}$ in the middle range. However, there are significant regional differences. More than 70% of the people living in NAWE live in cities with concentrations of less than 30µg/m$^3$, meeting the most stringent standards. In contrast, more than 70% of the populations in SEAR-D, WPR-B, EMR-D and SEAR-D live in cities where concentrations exceed even the most lenient standards.

This difference can also be seen in the estimated population-weighted mean concentrations of $PM_{10}$ for each subregion. These are computed from 1999 estimates of concentrations of $PM_{10}$ in cities, using the populations of each city in 2000 as weights. We have not directly used data for cities with data from monitoring in computing the subregional averages, to avoid incorporating short-term transitional variations into our exposure estimates.The mean exposures in the most polluted subregions are about three times higher than those in the least polluted subregions. Finer particles account for a larger share of PM in the highly industrialised countries of AMR-A and EURA compared to the other less industrialised subregions. We quantified the uncertainty in our estimates of the subregionspecific mean concentrations of PM using a bootstrap technique. In this method, the model is re-estimated many times (200 trials) using a randomly repeated sample of the observations used in estimating the model. For each trial, city and population-weighted subregional predictions of PM are generated using the methods described above. The degree of certainty in the point estimates of concentration of $PM_{10}$ for each subregion is directly related to the number of observations available from monitoring of PM. For example, the two subregions with the most frequently monitored cities also have the smallest confidence intervals for $PM_{10}$ values. In contrast, the five subregions with two or fewer cities that are monitored have larger con-fidence intervals than the other subregions. The width of the confidence intervals for these subregions depends on the geographic and climatic similarity of their cities with monitored data. For example, confidence intervals for AFR-E and EMR-B are about half of those for AFR-D and for EMR-D.

The estimates also show that substantial differences exist in the average concentration of PM within each subregion. The share of the urban population in cities with populations >100000 and in national capitals according to estimated concentrations of $PM_{10}$ is given in Figure 1. All cities in AMR-A, EUR-A, EUR-C and WPR-A are estimated to have concentrations of $PM_{10}$ of <60μg/m³ . In contrast, 95% of the urban population in SEAR-B and about 82% of the urban population in WPRB and EMR-D are exposed to >60μg/m³ $PM_{10}$. We also estimate that a high proportion of the urban population in SEAR-D is exposed to high annual average concentrations of $PM_{10}$.

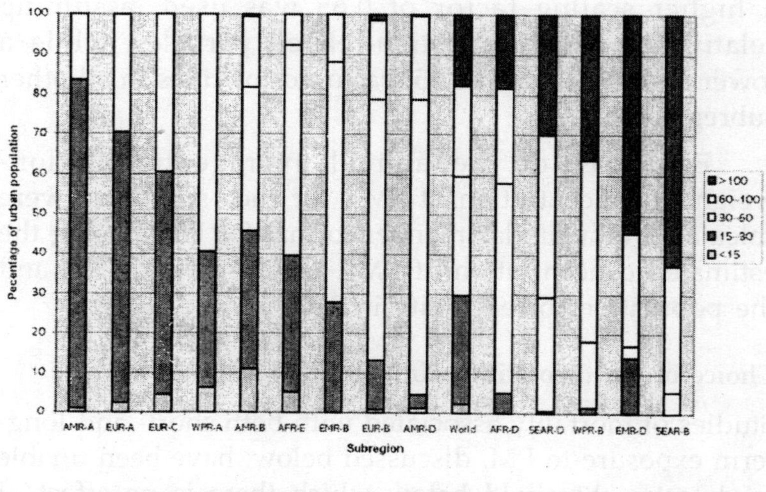


*Figure 1: Distribution of the urban population according to estimated concentrations of $PM_{10}$ in cities with populations of >100 and in national capitals, by subregion*

Since some of the health outcomes are based on $PM_{2.5}$,

rather than $PM_{10}$, concentrations for this pollutant had to be estimated. Cityspecific concentration of $PM_{2.5}$ was estimated as a fixed proportion of $PM_{10}$. Available measurements indicate that the ratio of $PM_{2.5}$ to $PM_{10}$ ranges from 0.5 to 0.8 in many urban areas in developed countries. Limited evidence suggests that a similar ratio may exist in large cities in other subregions. For example, a recent study from China reports the $PM_{2.5}$:$PM_{10}$ ratio to be in the range of 0.51 to 0.72 in four urban locations. However, in areas impacted by more crustal particles, the ratios are likely to be much lower. These areas will have a greater proportion of $PM_{10}$ in the coarse size range of 2.5–10mm. For example, evidence from the Coachella Valley (i.e. the Palm Springs area), an arid range of southern California suggests that the $PM_{2.5}$:$PM_{10}$ ratio is 0.35. Therefore, we assumed a ratio of 0.5 for our base case and have examined the sensitivity of our results to this assumption. Specifically, for our sensitivity analysis, for cities in AMR-A, EUR-A, EUR-B, EURC and WPR-A, a higher scaling factor of 0.65 was used, assuming relatively more combustion-related particles, while a lower scaling factor of 0.35 was used for cities in all other subregions.

Estimates of the annual average population-weighted concentration of $PM_{2.5}$ for each subregion were calculated in a similar manner to that for $PM_{10}$, using the estimated concentration of $PM_{2.5}$ for the city in 1999 and the population for each city in 2000.

### Choice of the theoretical-minimum-risk exposure

Studies of mortality associated with both short- and long-term exposure to PM, discussed below, have been unable to detect a threshold below which there is no effect of exposure. For most results presented below, we estimated the burden of disease with respect to a counterfactual concentration of 7.5μg/m³ $PM_{2.5}$ (or 15μg/m³ $PM_{10}$). This value is close to the lowest concentration observed in the epidemiological study from which we derived the concentration–response functions used for the majority of our estimates. This choice avoids extrapolating the concentration–response function(s) below the concentrations actually observed in the epidemiologic studies from which they were derived, although health benefits may well accrue from reductions below those concentrations.

We were aware, however, that for some cities the estimated concentrations of PM are lower, e.g. in AMR-A (United States and Canada), and that achieving such concentrations more widely would be not only desirable, but also feasible in some settings. Moreover, previous impact estimates have been sensitive to where this value was set. Therefore, we also conducted sensitivity analyses in which the theoretical minimum concentration was halved and doubled.

## Health Effects of Exposure to Urban Air Pollution

The past 10–15 years have seen a rapid increase in research on the health effects of air pollution, and it is now widely accepted that exposure to urban air pollution is associated with a broad range of acute and chronic health effects, ranging from minor physiological disturbances to death from respiratory and cardiovascular disease. Recently, a committee of the American Thoracic Society identified effects on respiratory health associated with air pollution, which should be considered adverse, spanning outcomes from death from respiratory diseases to reduced quality of life, and included some irreversible changes in physiological function. In general, the frequency of occurrence of the health outcome is inversely related to its severity, with the consequence that assessing total health impact solely in terms of the most severe, but less common, outcomes, such as mortality, will underestimate the total health burden of air pollution.

A large body of epidemiological research, discussed in more detail below, provides evidence that exposure to air pollution is associated with increased mortality and morbidity. The respiratory and cardiovascular systems appear to be the most affected. A growing body of toxicological and clinical evidence currently offers some limited insight into the mechanisms through which exposure to air pollution may produce the effects on respiratory and cardiovascular outcomes observed in epidemiological studies. These mechanisms may involve decrements in pulmonary func-tion, effects on heart rate variability and inflammatory response. Longterm bioassays and other studies of toxicity provide evidence for the mutagenicity and/or carcinogenicity of some components of urban air pollution, such as emissions from diesel-powered vehicles.

Air pollution may elicit both acute and chronic biological responses. Acute responses to air pollution in otherwise healthy persons may be confined to reversible physiological adaptations resulting from natural defence mechanisms. Acute responses may, however, also increase the severity or duration of an already established respiratory infection or of diseases such as asthma or chronic obstructive lung disease that have already placed the individual in a vulnerable position, and increase the risk of hospital admission or even death. If such vulnerability were temporary, for example, a severe infection of the lower respiratory tract, the individual might have recovered and lived for some time, had it not been for the added factor of exposure to air pollution at the time the individual was most vulnerable because of the infection. On the other hand, if the individual had a terminal chronic condition, such as severe chronic obstructive pulmonary disease or chronic congestive heart failure, exposure to air pollution might advance death by only a short time, this being imminent in any case. There is limited epidemiological evidence to suggest that ambient air pollution may contribute to the development of diseases such as chronic obstructive pulmonary disease, for which smoking and, in developing countries, indoor air pollution, are also risk factors.

Distributions of short-and long-term vulnerability, reflecting the prevalence of acute and chronic cardiorespiratory disease, may well differ across populations worldwide. This will have implications for the transferability of risk functions from studies in populations in NAWE to populations where differences in genetic factors, diet, tobacco smoking, extent of urbanisation, distribution of wealth and other factors related to social class, have resulted in different patterns of disease. For example, recent analyses of two cohorts in the United States showed clearly that the effects of long-term exposure to air pollution on mortality depend on attained educational level, with the largest relative effects observed among the least educated. Recent studies in developing countries have also reported such gradients in the relative risks of mortality. It is not clear what factor(s) might be responsible for these observations (e.g. aspects of occupation or diet), but it is reasonable to expect that they might vary across the globe. Differences in vulnerability to air pollution introduce a source of uncertainty in our estimates that can currently be only partially quantified.

Epidemiological evidence about exposure–response relationships is most directly applicable to the risk assessment of air pollution, because it comes from the direct observation of human populations under relevant conditions. Epidemiological studies have limitations that are largely a result of their observational nature. These relate to the accurate measurement of exposure, definition of outcomes and interpretation of associations that are observed. Assessing the causality of such associations requires a process of scientific reasoning that considers all evidence, including that from experimental studies. While there remain many gaps in our knowledge about the explanations for epidemiological associations, they can provide the best evidence to guide action to reduce the exposure of the population to air pollution, and to undertake health impact assessments, provided the uncertainties are recognised.

The epidemiology of air pollution takes advantage of the fact that concentrations of urban air pollution, and thus human exposure, vary in both time and space. For the most part, current epidemiological research has focused on either one or the other dimension, but infrequently on both within the same population(s). Short-term temporal variation in concentrations of air pollution over days and weeks has been used to estimate effects on daily mortality and morbidity. Spatial variation in long-term average concentrations of air pollution has provided the basis for cross-sectional and cohort studies of long-term exposure.

# 5

# Vehicular Air Pollution

Air pollution has been aggravated by developments that typically occur as countries become industrialised: growing cities, increasing traffic, rapid economic development and industrialisation, and higher levels of energy consumption. The high influx of population to urban areas, increase in consumption patterns and unplanned urban and industrial development have led to the problem of air pollution. Currently, in India, air pollution is widespread in urban areas where vehicles are the major con‑'butors and in a few other areas with a high concentration of industries and thermal power plants. Vehicular emissions are of particular concern since these are ground level sources and thus have the maximum impact on the general population. Also, vehicles contribute significantly to the total air pollution load in many urban areas.

## Pressure

### Increase Ease in Urban Population

Between 1951 and 1991, the urban population has tripled, from 62.4 million to 217.6 million, and its proportion has increased from 17.3% to 25.7%. Nearly two-thirds of the urban population is concentrated in 317 class I cities (population of over 100 000), half of which lives in 23 metropolitan areas with populations exceeding 1 million. The number of urban agglomerations/cities with populations of over a million has increased from 5 in 1951 to 9 in 1971 and 23 in 1991.

This rapid increase in urban population has resulted in unplanned urban development, increase in consumption patterns and higher demands for transport, energy, other infrastructure, thereby leading to pollution problems.

### Increase Ease in Number of Vehicles

The number of motor vehicles has increased from 0.3 million in 1951 to 37.2 million in 1997. Out of these, 32% are concentrated in 23 metropolitan cities. Delhi itself accounts for about 8% of the total registered vehicles and has more registered vehicles than those in the other three metros (Mumbai, Calcutta, and Chennai) taken together. Figure 1 shows the steep growth in the number of vehicles in India.

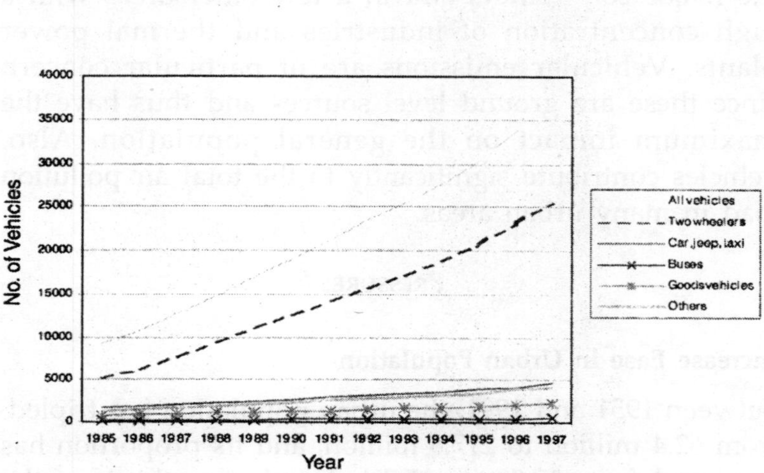


*Figure 1*

At the all-India level, the percentage of two-wheeled vehicles in the total number of motor vehicles increased from 9% in 1951 to 69% in 1997, and the share of buses declined from 11% to 1.3% during the same period. This clearly points to a tremendous increase in the share of personal transport vehicles. In 1997, personal transport vehicles constituted 78.5% of the total number of registered vehicles.

Road-based passenger transport has recorded very high growth in recent years especially since 1980-81. It is estimated that the roads accounted for 44.8 billion passenger kilometer (PKM) in 1951 which has since grown to 2,515 billion PKM in 1996. The freight traffic handled by road in 1996 was about 720 billion tonne kilometer (TKM) which has increased from 12.1 TKM in 1951. In contrast, the total road network has increased only 8 times from 0.4 million kms in 1950-51 to 3.3 million kms in 1995-96. The slow growth of road infrastructure and high growth of transport performance and number of vehicles all imply that Indian roads are reaching a saturation point in utilising the existing capacities. The consumption of gasoline and HSD has grown more than 3 times during the period 1980-1997. While the consumption of gasoline and HSD were 1,522 and 9,050 thousand tonnes in 1980-81, it increased to 4,955 and 30,357 thousand tonnes in 1996-97, respectively.

**Increase ease in Industrial Activity**

India has made rapid strides in industrialisation, and it is one of the ten most industrialised nations of the world. But this status, has brought with it unwanted and unanticipated consequences such as unplanned urbanisation, pollution and the risk of accidents.

The CPCB (Central Pollution Control Board) has identified seventeen categories of industries (large and medium scale) as significantly polluting and the list includes highly air polluting industries such as integrated iron and steel, thermal power plants, copper/zinc/ aluminium smelters, cement, oil refineries, petrochemicals, pesticides and fertiliser units.

The state-wise distribution of these pre-1991 industries indicates that the states of Maharashtra, Uttar Pradesh, Gujarat, Andhra Pradesh and Tamil Nadu have a large number of industries in these sectors. The categorywise distribution of these units reveals that sugar sector has the maximum number of industries, followed by pharmaceuticals, distillery, cement and fertiliser. It also indicates that agro-based and chemical industries have major shares of 47% and 37% of the total number of industries respectively. The status of pollution control as on 30 June 2000 is as follows: out of 1,551 industries, 1,324 have so far been provided the necessary pollution control facilities, 165 industries have been closed down and the remaining 62 industries are defaulters. It may be noted that in some of the key sectors such as iron and steel, 6 out of 8 units belong to the defaulters category in terms of having pollution control facilities to comply with the standards.

On the other hand, cement, petrochemicals and oil refinery sectors do not have any defaulters. Small scale industries are a special feature of the Indian economy and play an important role in pollution. India has over 3 million small scale units accounting for over 40 percent of the total industrial output in the country. In general, Indian small scale industries lack pollution control mechanisms. While the larger industries are better organised to adopt pollution control measures, the small scale sector is poorly equipped to handle this problem. They have a very high aggregate pollution potential. Also, in many urban centres, industrial units are located in densely populated areas, thereby affecting a large number of people.

**Increase Ease in Power Generation**

Since 1950-51, the electricity generation capacity in India has multiplied 55 times from a meager 1.7 thousand MW to 93.3 thousand MW. The generating capacity in India comprises a mix of hydro, thermal, and nuclear plants. Since the early seventies, the hydro-thermal capacity mix has changed significantly with the share of hydro in total capacity declining from 43% in 1970-71 to 24% in 1998-99. Thermal power constitutes about 74% of the total installed power generation capacity. However, increasing reliance on this source of energy leads to many environmental problems.

India's coal has a very high in ash content (24%–45%). The increased dependence of the power sector on an inferior quality coal has been associated with emissions from power plants in the form of particulate matter, toxic elements, fly ash, oxides of nitrogen, sulphur and carbon besides ash, which required vast stretches of land for disposal. During 1998-99, the power stations consumed 208 million tonnes of coal, which in turn produced 80 million tonnes of ash posing a major problem disposal.

Thermal power plants belong to the 17 categories of highly polluting industries. As on 30 June 2000, out of the 97 pre-1991 TPP's, 20 plants had not yet provided the requisite pollution control facilities.

## Domestic Pollution

Pollution from different types of cooking stoves using coal, fuelwood, and other biomass fuels contributes to some extent, to the overall pollution load in urban areas. For example, in Delhi, the share of the domestic sector is about 7%–8% of the total pollution load. The main concern is the use of inefficient and highly polluting fuels in the poorer households leading to a deterioration in indoor air-quality and health. However, a positive development in the domestic energy consumption is that liquefied petroleum gas is fast becoming the most popular cooking fuel, especially in urban areas, as it is cleaner and more efficient than traditional cooking fuels.

**Other Sources**

The problem of air pollution in urban areas is also aggravated due to inadequate power supply for industrial, commercial and residential activities due to, which consumers have to use diesel-based captive power generation units emitting high levels $NO_x$ and $SO_x$. In addition, non-point sources such as waste burning, construction activities, and roadside air borne dust due to vehicular movement also contribute to the total emission load.

***Air pollutant emission load***

The direct impact of a growth in various causal factors/ pressures is the increase in the emission loads of various pollutants, which has led to deterioration in the air quality. In India, there is no systemic time series data available related to air pollutant emission loads and trends. The availability of emission factors for Indian conditions is another issue that has not been given due attention so far.

TERI provides some broad estimates of the increase in pollution load from various sectors in India. The total estimated pollution load from the transport sector increased from 0.15 million tonnes in 1947 to 10.3 million tonnes in 1997. In 1997, CO claimed the largest share (43%) of the total, followed by $NO_x$ (30%), HC (20%), SPM (5%), and $SO_2$ (2%). Likewise, in the thermal power sector, the total estimated pollution load of SPM, $SO_2$ and $NO_x$ increased from 0.3 million tonnes in 1947 to 15 million tonnes in 1997. In 1997, SPM claimed the largest share (86%) of the total. In the industrial sector, the total estimated emissions of SPM from the 7 critical industries (iron and steel, cement, sugar, fertilisers, paper and paper board, copper and aluminium) increased from 0.2 million tonnes in 1947 to 3 million tonnes in 1997.

The World Bank study shows that pollution is concentrated among a few industrial sub-sectors and that a sector's contribution to pollution is often disproportionate to its contribution to industrial output. For example, petroleum refineries, textiles, pulp and paper, and industrial chemicals produce 27% of the industrial output but contribute 87% of sulphur emissions and 70% of nitrogen emissions from the industrial sector. Likewise, iron and steel, and non-metallic mineral products, produce about 16% of the industrial output but account for 55% of the particulate emissions.

### *Vehicular emissions*

The drastic increase in number of vehicles has also resulted in a significant increase in the emission load of various pollutants.

The quantum of vehicular pollutants emitted is highest in Delhi followed by Mumbai, Bangalore, Calcutta and Ahmedabad. Carbon monoxide (CO) and hydrocarbons (HC) account for 64% and 23%, respectively, of the total emission load due to vehicles in all these cities considered together.

Apart from the concentration of vehicles in urban areas, other reasons for increasing vehicular pollution are the types of engines used, age of vehicles, congested traffic, poor road conditions, and outdated automotive technologies and traffic management systems.

Vehicles are a major source of pollutants in metropolitan cities. In Delhi, the daily pollution load has increased from 1,450 tonnes in 1991 to 3,000 metric tonnes in 1997. The share of the transport sector has increased from 64% to 67% during the same period while that of the industrial sector (including power plants) has decreased from 29% to 25%.

*Ambient air quality*

Under the National Ambient Air Quality Monitoring (NAAQM) network, three criteria air pollutants, namely, SPM, $SO_2$, and $NO_2$ have been identified for regular monitoring at all the 290 stations spread across the country.

CPCB analyses the status and trends of air quality at various cities in India for the period 1990-98. The most prevalent form of air pollution appears to be SPM although there are many stations at which $SO_2$ and $NO_2$ levels exceed permissible limits. The high influx of population to urban areas increase in consumption patterns, unplanned urban and industrial development and poor enforcement mechanism has led to the problem of air pollution.

The government has taken a number of measures such as legislation, emission standards for industries, guidelines for siting of industries, environmental audit, EIA, vehicular pollution control measures, pollution prevention technologies, action plan for problem areas, development of environmental standards, and promotion of environmental awareness. However, despite all these measures, air pollution still remains one of the major environmental problems. At the same time, there have been success stories as well such as the reduction of ambient lead levels and comparatively lower $SO_2$ levels.

**SPM**

Suspended particulate matter is one of the most critical air pollutants in most of the urban areas in the country and permissible standards are frequently violated several moni-tored locations. Its levels have been consistently high in various cities over the past several years.

The annual average minimum and maximum SPM concentration in residential areas of various cities ranged from 60 μg/m³ to 521 μg/m³, while in industrial areas the annual average ranged between 53 μg/m³ and 640 μg/m³. The mean of average values of SPM for nine years ranged between 99 μg/m³ and 390 μg/m³ in residential areas and between 123 μg/m³ and 457 μg/m³ in industrial areas indicating that the annual average limit of suspended particulate matter for residential areas (140 μg/m³) and for industrial areas (360 μg/m³) had been frequently violated in most cities.

The maximum suspended particulate matter (SPM) values were observed in Kanpur, Calcutta, and Delhi, while low values have been recorded in the south Indian cities of Chennai, Bangalore, and Hyderabad. The SPM non-attainment areas are dispersed throughout the country. The states with maximum SPM problems are Gujarat, Maharashtra, and Madhya Pradesh, where SPM problems are high to critical in a large number of cities. The widespread criticality of the SPM problem in the country is due to the synergistic effects of both anthropogenic and natural sources. Some of these are extensive urbanisation and construction activities, vehicular pollution increase, extensive use of fossil fuel in industrial activities, inadequacy of pollution control measures, biomass burning, presence of large acid and semi-acid area in north-west part of India, increasing desertification, and decreasing vegetation cover.

## $SO_2$

The annual average level fluctuation of $SO_2$ was highest in residential areas of Howrah (West Bengal) recording between 40.6 μg/m³ and 103.8 μg/m³ while it was quite low in residential areas of Nagpur, Chandigarh, and

Jaipur (below 10 μg/$m^3$). Among the industrial areas, the recorded sulphur dioxide levels were high at Pondicherry, Calcutta, Mumbai, and Howrah, and low at Nagpur, Jaipur, and Chandigarh. Thus, based on the mean average sulphur dioxide value, Nagpur, Chandigarh and Jaipur are cities with the least problems related to sulphur dioxide in the ambient air, while the problem is significant in Howrah, Calcutta, and Pondicherry, where annual average limits (60 and 80 μg/$m^3$ for residential and industrial areas) have been violated many times during the past several years.

The sulphur dioxide levels have generally attained air quality standards in the country except some cities of dense urban and industrial activities like Dhanbad (Bihar), Ahmedabad, Ankleshwar, Vadodara and Surat (Gujarat); Nagda (Madhya Pradesh); Pondicherry; and Howrah (West Bengal). Some of the measures taken such as cleaner fuel quality (reduction of sulphur content in diesel) and switch over to cleaner fuel option have contributed t lower $SO_2$ ambient levels.

**$NO_2$**

The air quality monitoring data indicate that the annual average nitrogen dioxide has been well within the annual average limit (60 μg/$m^3$ for residential area and 80 μg/$m^3$ for industrial areas) at most urban cities except in some years in residential areas of Howrah, Vishakhapatnam, Kota, and industrial areas of Howrah. The annual average concentration has been low at Nagpur, Chennai, Kanpur, and Chandigarh while levels are moderate in other cities. The nitrogen dioxide non-attainment areas were at Vishakhapatnam (AP), Jabalpur (MP), Pondicherry, Alwar, Kota, Udaipur (Rajasthan) and Howrah (West Bengal). The criticality of problem was observed at Vishakhapatnam, Kota, and Howrah.

**Status of Air Pollutants**

The salient results of additional parameters at some stations in the metropolitan cities of Delhi, Calcutta, Mumbai, and Chennai for the years 1996-97 are as follows:

— The annual mean concentration of respirable particulate matter (RPM) is much higher than the prescribed limits of 120 μg/ $m^3$ (industrial) and 60 μg/$m^3$ (residential and other uses) in Delhi and Calcutta. However, RPM in Mumbai and Chennai is not very high but is greater than prescribed ambient air quality standards.

— Though particulate lead in the ambient air of Calcutta and Delhi is higher as compared to the other two cities (Chennai and Mumbai), it is well within the prescribed limits for the different area classes.

— The concentration of polycyclic aromatic hydrocarbons (PAHs) is showing an upward trend. However, at present permissible limits for PAHs have not been notified.

It may be noted that lead free gasoline has been introduced throughout the country w.e.f. February 1, 2000, and in all metropolitan cities since 1995. This has resulted in a downward trend in the lead concentrations in the ambient air.

## Air Quality at Traffic Intersections

Air quality monitoring conducted at different traffic intersections in Delhi revealed the following:

— Respirable particulate matter was excessively high at all the monitoring locations.

— Sulphur dioxide was recorded within limits at all the locations.

- Nitrogen dioxide was recorded well within the limits except a few locations.
- The carbon monoxide levels at most locations was much higher than the prescribed permissible limit. This is because of high traffic density and large number of motor vehicles operating on the roads.

Measures such as stringent emission norms for vehicles, cleaner fuel quality, inspection and maintenance programmes are expected to make some contribution towards improvement in the air quality. However, in the absence of mass transport system, the tremendous increase in personal vehicles is a cause for concern.

## Air Pollution and Health Impacts

In India, millions of people breathe air with high concentrations of dreaded pollutants. The air is highly polluted in terms of suspended particulate matter in most cities. This has led to a greater incidence of associated health effects on the population manifested in the form of sub-clinical effects, impaired pulmonary functions, use of medication, reduced physical performance, frequent medical consultations and hospital admissions with complicated morbidity and even death in the exposed population. As per a World Bank study, respiratory infections contribute to 10.9% of the total burden of diseases, which may be both due to presence of communicable diseases as well as high air pollution levels, while cerebro vascular disease (2.1%) ischemic heart disease (2.8%) and pulmonary obstructions (0.6%) are much lower. The prevalence of cancer is about 4.1% amongst all the diseases indicating that the effects of air pollution are visualised on the urban population .

A WHO/UNEP study compared standardised prevalences of respiratory diseases in different areas of Mumbai, classified according to ambient average concentrations of sulphur dioxide. The study revealed a relatively higher prevalence of most respiratory diseases in polluted urban areas than in the rural control area.

In India, in a study of 2031 children and adults in five mega cities, of the 1852 children tested, 51.4% had blood lead levels above 10 µg/dl. The percentage of children having 10 µg/dl or higher blood lead levels ranged from 39.9% in Bangalore to 61.8% in Mumbai. Among the adults, 40.2% had blood lead levels of about 10 µg/dl.

Brandon, Hommann, and Kishor estimated the total magnitude of economic costs associated with environmental degradation in India. Using the 1991-92 air pollution data for particulates, $SO_2$, $NO_X$, and lead from 36 cities, health impacts were estimated in terms of reductions in morbidity and mortality if pollutant levels in these cities were reduced to the WHO annual average standard. The total health costs due to air pollution were estimated to be $517-2102 million. Also, the physical impacts were in terms of 40,000 premature deaths avoided.

TERI estimated the incidence of mortality and morbidity in different groups in India due to exposure to $PM_{10}$ and translated these impacts into economic values. The results indicated 2.5 million premature deaths and total morbidity and mortality costs of Rs 885 billion to Rs 4250 billion annually.

## Noise Pollution

Studies by the CPCB on the ambient noise levels show that they exceed the prescribed standards in most of the big cities. The major sources of noise are vehicles and industrial manufacturing processes.

## Policy Response

### *Legislation*

The government has formulated a number of legislations, policies, and programmes for protecting the environment. Some of these related to air pollution are the Air (Prevention and control of pollution) Act, 1981 and the Environment (Protection) Act, 1986. India has also adopted the Male declaration on Control and Prevention of air pollution and its likely transboundary effects for South Asia in April 1998.

### *Ambient air quality standards*

Ambient air quality standards (both shortterm, i.e., 24 hourly, and long-term, i.e., annual) have been laid down for industrial, residential/rural/other, and sensitive areas with respect to pollutants such as $SO_2$, $NO_X$, SPM, RPM, Pb, and CO. Ambient air quality standards for NH3 have also been notified.

Guidelines for siting of industries Guidelines for siting industries are prescribed so that the possible adverse effects on the environment and quality of life can be minimised. Some natural life-sustaining systems and specific land-uses are more sensitive, and that has been taken into account while specifying the minimum prescribed distance for siting a given industry.

### *Environmental impact assessment (EIA)*

EIA is mandatory for 29 specific activities/ projects and also for some of the activities to be taken up in identified areas such as the coastal zone, Doon valley, etc. The procedure for examining the impact of different activities includes the preparation of EIA report, holding of a public hearing and examination by a duly constituted expert committee. Also, MoEF has taken up carrying capacity-based regional planning studies in certain selected areas of the country.

***Emission standards for industries***

The CPCB has laid down the maximum permissible limits for different pollutants for many categories of industries that contribute to air pollution. The standards have been notified by MoEF under the Environment (Protection) Act 1986.

***Environmental audit***

Submission of an environmental statement by polluting units to the concerned State Pollution Control Boards has been made mandatory under the Environment (Protection) Act, 1986.

Zoning atlas for siting of industries In order to delineate the areas that are suitable for industrial siting, a district-wise zoning atlas project has been taken up by the CPCB that zones and classifies the environment in a district. The industrial zones are identified based on the sensitivity and pollution-receiving potential of the district.

***Pollution Prevention Technologies***

Industries are encouraged to use cleaner and low waste or no waste technologies to reduce waste generation and the emission of pollutants. Various pollution prevention technologies are being developed and promoted. A good scope exists for the demonstration and replication of cleaner technologies in clusters of small-scale industries such as foundry (Howrah), pottery (Khurja), and glass (Firozabad).

***Beneficiated coal***

The Ministry of Environment and Forests has made it

mandatory for thermal power plants located beyond 1000 km from the coal pithead, or in urban, ecologically sensitive or critically-polluted areas, to use beneficiated/ blended coal containing ash not more than 34%, with effect from June 2001. The power plants using FBC (Fluidized Bed Combustion) and IGCC (Integrated Gasification Combined Cycle) combustion technologies are, however, exempted to use beneficiated coal irrespective of their locations.

***Pollution control in problem areas***

Twenty-four problem areas have been identified in the country for pollution control through concerted efforts involving all the concerned agencies / industries. Action plans have been prepared and are being implemented.

## Control of Vehicular Pollution

The various measures taken by government to mitigate emissions from transport sector are as follows:

*Stringent emission norms.* The mass emission standards for new vehicles had been first notified in the year 1991 in India. Stringent emission norms along with fuel quality specifications were laid down in 1996 and 2000.

Euro I norms are applicable from 1 April 2000 and Euro II norms will be applicable all over India from 1 April 2005. However, in the case of the NCR, the norms were brought forward to 1 June 1999 and 1 April 2000 for Euro I and Euro II, respectively.

*Cleaner fuel quality.* To conform to the stringent emission norms, it is imperative that both fuel specification and engine technologies go hand in hand. Fuel quality specifications have been laid down by the BIS (Bureau of Indian Standards) for gasoline and diesel for the period 2000-2005 and beyond 2005 for the country.

Given the increased usage of diesel in our country, it becomes necessary to reduce its sulphur content. In a recent directive by the Supreme Court, the Ministry of Petroleum and Natural Gas is to supply diesel with 0.05% m/m sulphur to the NCT by 31 December 2000 and entire NCR from 30 June 2001. For gasoline, lead has been phased out in the entire country w.e.f. 1 February 2000. Similarly the benzene content is to be reduced and by 1 October 2000, gasoline with 1% benzene is to be supplied to the whole of the NCT region. For NCR, it should be supplied by 31 March 2001. Later, it has to be extended to other parts of the country as well.

Inspection and maintenance (I&M). The first and most important step towards emission control for the large in-use fleet of vehicles is the formulation of an inspection and maintenance system. It is possible to reduce 30-40% pollution loads generated by vehicles through proper periodical inspections and maintenance of vehicles. I&M measures for in-use vehicles are an essential complement to emission standards for new vehicles. In India, the existing mechanism of I&M is inadequate. Thus, there is a great need to establish effective periodic I&M programmes.

Other stringent measures in certain areas. On1st April 1999, the specifications for 2T oil became effective. In order to prevent the use of 2T oil in excess of the required quantity, premixed 2T oil dispensers have been installed in all gasoline stations of Delhi. Other measures include bans on commercial vehicles more than 15 years old, a ban on the registration of new auto-rickshaws with front engine, replacement of all pre-1990 autos and taxis with new vehicles using clean fuels; and the removal of 8 year old buses from the roads unless they use CNG or some other clean fuel. It is also planned that all buses in Delhi are to switch over to CNG instead of diesel by 31 March 2001.

*Role of the judiciary*

In recent years, the judiciary has played a prominent role in environmental protection. A number of judgements relating to stringent vehicle emission norms, fuel quality, introduction of cleaner fuels, phasing-out of older vehicles, and shifting of hazardous industries have provided a great deal of momentum to the efforts for improvement of air quality.

**Policy Gaps**

— Prevention based environmental policy needs to be strengthened. Issues such as cleaner technology and land use planning incorporating environmental considerations need to be given priority.
— Effectiveness and impact of various policy measures not assessed.
— No separate transport policy exists at the national and state levels.
— No well defined policy to promote private participation in public transport.
— Lack of coordination between various government agencies to improve transport services.

**Knowledge/infor information/mation/data gaps**

— Strengthening of monitoring at hotspots/ traffic intersections; more stations to be established and frequency of monitoring increased.
— Additional air quality parameters need to be monitored such as ozone, benzene, PAH, $PM_{2.5}$, dry deposition of sulphates and nitrates.
— Private/Community participation in monitoring activity.
- Emission factor development for various activities.

— Emission load mapping at regular intervals for all the urban areas.

— Air pollution modelling as a tool for forecasting and urban planning.

— Strengthening of information on number of vehicles on road, vehicle usage, etc.

**Policy Recommendations**

Despite the aforementioned legislative/policy measures as well as a host of other decisions taken by the government, air pollution remains a major concern. Besides continuing and consolidating the ongoing schemes/ programmes, new initiatives and definite programmes need to be formulated for the efficient management of urban air pollution.

***Vehicle pollution control***

Since vehicles contribute significantly to the total air pollution load in most urban areas, vehicular pollution control deserves top priority. A practical strategy should be devised that reduces both emissions and congestion, using a mixed set of instruments, which are dictated by command and control, and/or the market based principles. Some of the these are:

— Augmentation of public transport system.

— Mass Rapid Transport System may be considered for the fast expanding and major urban areas in the country.

— Incentives and regulations affecting vehicles with a view to reducing the rate of growth in ownership of personal vehicles.

— Traffic planning and management. Also, construction of express highways linking major urban areas should be undertaken.

- Taxes on fuels, vehicles—the revenue so generated could be used for pollution control measures.
- Further tightening of emission norms and fuel quality specifications.
- Greater promotion and use of alternative fuels such as CNG/LPG/Propane/ battery operated vehicles. Expansion of CNG dispensing facilities and increased fiscal incentives for CNG kits.
- Replacement of two-stroke engines.
- Curbing fuel adulteration—state-of-the-art testing facilities and deterrent legal action.
- Strengthening of inspection and maintenance (I&M) system: The I&M system, comprising inspection, maintenance, and certification of vehicles, is crucial for regulating pollution for the large fleet of in-use vehicles. It should include testing of various elements of safety, road worthiness and compliance to pollution norms.

*Industrial pollution control*

- Thrust for cleaner technologies
  - Waste minimisation technologies involving process change, raw material substitution, improved housekeeping, etc.
  - Waste utilisation technologies involving reclamation and utilisation of wastes as secondary raw materialFlue gas desulphurisation
  - Combustion modification for $NO_x$ reduction
  - Incentives for the development and adoption of clean technology and emission reduction
- Database on clean technology
  - Database on available technologies, their

performance, sources, investment required, etc, should be created, regularly updated, and widely disseminated

- Strengthening of emission standards
  - Emission standards for various categories of industries need to be strengthened. To shift from pollution control to pollution prevention, rules related to load based standards instead of concentration based standards need to be enforced.
- Appropriate siting of high pollution potential industries/projects
- Fiscal incentives for pollution prevention and control measures

**Other Pollution Abatement Measures**

- Strengthening of monitoring network
  - The monitoring network requires a massive quality control programme and expansion of its operations to cover new stations as well as more pollutants on a regular basis. Smaller cities should also be covered so that preventive measures could be taken before the pollution problem becomes acute.
- Information dissemination/mass awareness/ training
  - State-of-the-art technology should be used for wider dissemination of environmental information. Transparency and access to the data to be improved. Measures such as pollution bulletins and air pollution forecasts should be started on a regular basis
  - Massive thrust should be provided to mass awareness campaigns involving community

organisations such as residents associations, students, voluntary bodies and NGOs. Strategic action plans for implementation should be devised

- Support measures such as training and education for the industry, governmental agencies, and the public, as well as greater coordination among institutions, are important

— Air quality management strategy

- A comprehensive urban air quality management strategy should be formulated that includes information related to urban planning, ambient air quality, emission inventory, and air quality dispersion models.
- Effectiveness of EIA as a tool and environmental audit needs to be critically assessed
- Systematically planned emission load mapping studies should be undertaken at regular intervals. Development of emission factors for Indian conditions should be taken up

— Fiscal measures

- Economic instruments need to be in place to encourage a shift from curative to preventive measures, internalisation of the cost of environmental degradation, and conservation of resources. The revenue generated may be used for enforcement, collection, treatment facilities, and research and development.
- Incentives for environmentally benign substitute, technologies and energy conservation

— Promotion of renewable energy sources such as hydro, wind, and solar

— Use of cleaner fuels like LPG and kerosene for

domestic consumption would reduce indoor air pollution

- Air quality standards should be based on local dose-response relationships for which appropriate environmental epidemiological studies should be undertaken
- Non-point sources of pollution also to be controlled such as pollution from generators, waste burning, etc
- Increase in green cover. Appropriate design of green belts/barriers and proper selection of plant species
- Noise pollution is also a major problem in metro cities and adequate preventive and control measures need to be taken.
- Enforcement mechanism
  - Significant improvements in the enforcement mechanism required to ensure that the policies are implemented both in letter and spirit
  - Wherever necessary, the policies/standards need to be reformulated keeping in mind the fast-changing scenario
  - An effective environment management plan should be devised that includes environmental strategy, regulation, institutional capacity-building, and economic incentives and penalties.

# 6

# Air Pollution, Ozone and Aerosols

The effects of anthropogenic aerosols on the climate constitute one of the largest sources of uncertainty in quantifying climate change. Air pollution in Asian cities has grown with the progressing industrialisation and urbanisation. This recent experience in Asia is predated by similar problems in the western countries at early stages of their economic development. By studying how coal burning produced the London smog problem and how the use of petroleum led to photochemical smog in Los Angeles we can anticipate the types of air pollution problems that can be expected in Asian cities as well.

## POLLUTION PROBLEMS

Studies from both London and Los Angeles help to forecast how serious pollution problems elsewhere eventually will be solved. Over the years, great improvements have been made in the air quality in London and Los Angeles.

### Use of Coal

Wood was used for fuel in London up to the 13$^{th}$ century when wood supplies became depleted. By the 1280's, coal was used for fuel in industrial processes including limekilns and metalworking. Coal use in these industries produced smoke concentrations that merited control, sometimes through relocating the industry to the downwind side of town.

By the 1600's, coal became widely used for domestic heating in London in addition to expanded industrial use as the city grew. Concern was registered by local officials due to the soiling of buildings by black soot, including St. Paul's Cathedral. Effects on public health also were debated. In the 1700's, concern was raised about the sulfurous content of urban air. Black smoke and sulfur oxides both are produced when coal is burned, and it is often difficult to attribute ill effects to either pollutant separately as they tend to travel together. Persons adversely affected by the urban air pollution levels were urged to leave the city in order to improve their health. Indications are that smoke concentrations were so heavy in the 1700's in London that some buildings had to be repainted every three years to hide the effects of smoke particle deposition. The 19th century saw continued growth in coal use in London as population shifted to the city in search of the expanded job opportunities to be found there. Smoke abatement laws were passed but there was little effective control because the technology for control was lacking. Increasing coal use was accompanied by an increasing frequency of fog in the city. Sulfuric acid droplets and particulate sulfates derived from coal combustion effluent as it ages in the atmosphere provide hygroscopic sites for fog droplet nucleation making it easier for fogs to form. The word smog follows from this combination of circumstances, as it is a contraction of the words smoke and fog.

The use of coal and the frequency of fogs both peaked in London in about the year 1900. Upgraded furnaces and use of automatic rather than hand stoking helped to reduce the amount of smoke produced when burning coal. At the turn of the century, London experienced more than 60 foggy days per year, dropping to about 20 days per year in 1950. During this period, it became obvious that increased death rates were associated with the heavily polluted London fogs. In 1952 an extended period of air stagnation lead to an episode of polluted fog that resulted in 4000 excess deaths in London over a several day period. Today the atmosphere in London is much cleaner. Domestic and small industrial use of coal has been discontinued in favor of use of clean burning natural gas and electric rather than steam driven motors and engines.

**Photochemical Smog**

The air pollution problem in Los Angeles grew to unacceptable levels during the period of rapid industrialisation that accompanied the Second World War. The most obvious characteristics of this new air pollution problem included severe eye irritation and greatly reduced visibility. The local visibility problem was so severe that there was almost a complete loss of days with extremely clear air. During the period 1932-1937, very clear afternoons with visual range in excess of 35 miles occurred on 21% of the days of the summer season. By 1943-1947, the number of such very clear afternoons dropped to 0.2% of summer days.

In the early 1950's Professor Haagen-Smit of the California Institute of Technology demonstrated that a key feature of Los Angeles smog involved photochemical reactions that occur in the atmosphere when light shines on a mixture of hydrocarbon vapors and oxides of nitrogen. These reactions lead to ozone formation, as well as to the formation of a large variety of co-pollutants including aldehydes, peroxyacyl nitrates (PANs), and fine particles produced from the low vapor pressure products of atmospheric chemical reactions. Ozone is a respiratory irritant and can damage crops, trees and materials such as natural rubber. The aldehydes are eye irritants and could be responsible for much of the reported eye irritation.

Following the discovery of the general nature of photochemical smog, a network of air monitoring stations was established throughout the Los Angeles area. In the late 1950's, peak one-hour average ozone concentrations above the California ozone air quality standard of 0.1 ppm occurred on more than 300 days per year. Investigation of the oxides of nitrogen emissions to the Los Angeles area atmosphere showed emissions of about 400 tons per day in the late 1940's about half due to motor vehicle exhaust and half due to stationary source fuel use such as at electric utility boilers and petroleum refineries. Hydrocarbon vapor emissions at that time were much higher, more than 1500 tons per day. The largest hydrocarbon source initially was from evaporative losses or leaks from petroleum production, refining, and marketing. Los Angeles contains a number of oil fields and also the petroleum refineries needed to supply all of southern California and parts of Nevada and Arizona. Motor vehicle exhaust contributed the next largest hydrocarbon source, followed by evaporation of solvents used in cleaning, printing, and painting to name but a few of the many sources of solvent loss.

Stringent emission control rules were adopted, many of them directed toward stopping hydrocarbon losses in the petroleum industry. From the 1940's through the mid 1960's there was a large and steady decline in hydrocarbon emissions from the petroleum sector of the economy. At the same time, the population in the air basin surrounding Los Angeles grew from roughly 3 million persons in 1940 to nearly 9 million persons in the mid-1960's. With that increase in population came an increase in the number of automobiles. Hydrocarbon emissions from the expanded motor vehicle fleet grew to more than compensate for the decreased emissions from the petroleum production, refining and marketing sector. Reduced hydrocarbon emissions from cars, and especially the introduction of hydrocarbon oxidation catalysts on cars in 1975 later acted to reduce motor vehicle emissions. Meanwhile, population growth within the air basin surrounding Los Angeles continued, reaching about 14 million persons by the turn of the present century. With that growth came increased emissions of solvents from a variety of small uses in the community such as household cleaning products and paints. Today, the hydrocarbon emissions in Los Angeles are about evenly split between vehicle exhaust and evaporative sources.

Strict emissions controls changed the nature of the hydrocarbon emissions in the Los Angeles area, but the quantity of hydrocarbons emitted daily stayed at high levels of more than 1000 tons per day due to the steadily expanding population within the air basin that surrounds Los Angeles. Close to the source, ozone concentrations depend on the ratio of the various pollutants in the atmosphere more than they depend on the absolute amounts emitted. As population growth continued in the Los Angeles area, oxides of nitrogen emissions from fuel combustion increased from the level of 400 tons per day in the late 1940's to approximately 1600 tons per day by the mid 1970's. Use of 3-way catalysts on newer motor vehicles and application of other nitrogen oxides emissions controls to boilers and heaters subsequently lowered nitrogen oxide emissions in the air basin surrounding Los Angeles to approximately 1000 tons per day. Analysis of the Los Angeles ozone problem shows that reduction of the ratio of hydrocarbon emissions to oxides of nitrogen emissions is the easiest way to reduce ozone concentrations in that city. The increased oxides of nitrogen emissions relative to the 1940's combined with a strict hydrocarbon control program reduced the hydrocarbon to oxides of nitrogen ratio in the atmosphere, and a general reduction in emissions density occurred as the population spread over a much larger geographic area over time, resulting in steadily declining peak ozone concentrations over time. By 1990, the number of days per year with peak 1-hr average ozone concentration greater than the Federal air quality standard of 0.12 ppm had declined to roughly 160 per year. That downward trend has continued such that today far less than 100 days per year exceed 0.12 ppm ozone, and days with ozone concentrations above 0.2 ppm are rare indeed.

### Air Quality

Air quality in Chinese cities today more closely resembles the London smog problem than the Los Angeles smog problem, although that could change as present problems with coal smoke are brought under control. Coal is burned for power generation, industrial applications, in small domestic and commercial boilers and as a home heating fuel. Outside of a few major cities such as Beijing and Shanghai, motor vehicle populations and hence petroleum use is small when compared with American cities. Chinese cities experience very high airborne particle concentrations due to airborne dust, primary particles emitted from coal and biomass combustion, motor vehicle exhaust as well as secondary sulfates formed by atmospheric chemical reaction from the sulfur dioxide emitted when coal is burned. Fine particles smaller than 2.5 micrometers in diameter are responsible for most visibility problems and are of health concern because they can penetrate deep into the lung. Recent measurements in downtown Beijing show fine particle concentrations averaging just over 100 micrograms per cubic meter of air compared to about 22 micrograms per cubic meter annual average in Los Angeles. The annual average fine particle air quality standard proposed for the United States is set at 15 micrograms per cubic meter, about one seventh of the level in Beijing. Fine particle concentrations in Beijing approach 300 micrograms per cubic meter on the highest concentration days of the year, and these high concentration events often co-occur with fogs, as in London. Total suspended particulate matter concentrations (airborne particles of all sizes) can reach or exceed one milligram per cubic meter of air on a bad day. Officials in Beijing are taking aggressive steps to correct this air pollution problem. Natural gas has been brought into the city and is systematically being used to displace coal combustion. Use of natural gas instead of coal was a major feature of the successful cleanup of the London-smog problem.

New cars are now equipped with catalytic converters to reduce their exhaust emissions and unleaded gasoline is available for these cars. Catalytic converters are being retrofitted onto precatalyst cars built since the mid-1990's. Construction dust suppression measures are in use. Nevertheless, Chinese cities face a difficult problem when it comes to air pollution control. Airborne particle measurements made at background sites in the countryside outside of Beijing show fine particle concentrations more than half as high as in the city. This implies that much of the air pollution in Beijing is already in the air before the air enters the city. Indications are that there is a large-scale regional air pollution problem of great geographic extent covering North China that could frustrate control programs if those control programs are merely local in scope.

The South Asian Region includes the Indian subcontinent (India, Pakistan, Bangladesh, Nepal and Bhutan) as well as Sri Lanka, and Maldives. The region is one of the most densely populated in the world, with present population densities of 100-500 persons/km$^2$.

Although total land area comprises only about 3% of the world's land masses, the 1990 population was about 1/5 th of the global total and by 2025 is expected to rise to about – (Table 1). Three fourths of the region's population live in rural area, of which about one third is living at the threshold of poverty. Deforestation is serious in most of the countries within the region, although in India afforestation efforts have shown some positive results in recent years. Urban population is increasing rapidly and is expected to be nearly 50% in 2025 causing destruction of fragile ecosystem and increasing air pollution. The gross domestic product (GDP) growth rates are high, around 5-6%. There is increasing water stress: per capita availability of water has decreased drastically from 5200 $m^3$/cap/yr in 1950, to a far lower value of 1860 $m^3$/cap/yr in the year 2000 for India and from 5140 $m^3$/cap/yr in 1950 to 1200 $m^3$/cap/yr in Pakistan, that is below water stress level of 1700 $m^3$/cap/yr in the latter case. There are 5 megacities with population around or above 10 million in the region: Calcutta, Mumbai, Delhi, Dhaka and Karachi.

*Table 1: Population in South Asian Countries (in millions)*

| *Year* | *India* | *Pakistan* | *Bangladesh* | *Sri Lanka* | *Nepal* |
|---|---|---|---|---|---|
| 1990 | 853 | 123 | 112 | 17 | 19 |
| 1995 | 929 | 130 | 120 | 18 | 22 |
| 2025 | 1,442 | 267 | 235 | 25 | 35 |

**Climate**

The region is characterised by a tropical monsoon climate. Differences in rainfall are of primary significance in defining the climate of the region. The most important feature is the seasonal alteration of atmosphere flow patterns associated with the monsoon. Two monsoon systems operate in the region: the Southwest or summer monsoon (June-September) and the Northeast or winter monsoon (December-April). The rainfall during the summer monsoon largely accounts for the total annual rainfall over most of South Asia (except over Sri Lanka where rainfall of the winter (Northeast) monsoon is dominant) and forms a chief source of water for agriculture and other activities. The monsoon rainfall in South Asia is characterised by large spatial and temporal variability. The arid, semi-arid region encompassing Pakistan and Northwest India receive monsoon rainfall as low as 50 mm while parts of Northeast India and the west coast receive over 1000 mm. This region also features large year-to-year variations in the rainfall frequently causing severe floods/droughts over large areas.

There are two major anomalous regions: the arid and semi-arid parts comprising of large areas of Pakistan and north-western Indian states of Rajasthan, Punjab, Haryana and Gujrat which experience frequent droughts, and the eastern Himalayan sub-region, fed by the Ganga-Brahmaputra-Meghna river system, which are subjected to frequent floods. In India, during the period 1871-2000 there were 22 drought years and 19 flood years. There had been three cases of prolonged drought condition, viz., 1904-05, 1965-66 and 1985-87. Such cases cause great calamity. Similarly, there had been two cases of prolonged flood conditions, viz., 1892-94 and 1916-17. Studies indicate a clear relationship between the occurrence of droughts (floods) in South Asia with the El Niño (La Niña) events in the east Pacific Ocean. It has been observed that, during the period 1856-1997 there were 30 El Niño years in which the averaged monsoon rainfall over India was 7% below normal; in 10 out of these 30 cases, drought conditions prevailed over India. Two years featured flood conditions. During the same period there were 16 La Niña years, 9 of which featured flood conditions over India. However, it appears from some of the recent studies that this relationship has been weakening in recent years, possibly due to global warming.

The key parameters for economic development in the region are food and water supply. Both are under stress, especially in the two regions experiencing extreme climates, and this will increase in the future. While India is currently producing adequate food grains, the projected requirement of 330 MT in 2025 will not be easy to achieve. Similarly for Pakistan and Bangladesh, the large future requirements will be difficult to meet.

**Fossil Fuel and Biomass Burning**

While per capita emissions of greenhouse gases are very low for all countries in this region, the growth rate is high (about 4-6% per year). India is the only country heavily dependent on coal for energy in the region. Bangladesh uses primarily natural gas, Pakistan oil and natural gas, and Nepal primarily biomass fuel. Biofuel consumption is also large for India, but the estimates vary widely. Rough estimates are around 150-250 MT fuelwood, 90-100 MT dungcake and 40-100 MT agricultural residue. Bio fuels, therefore, could be around one-half of the total fuel for India. The total bio fuel for the entire subcontinent could be anywhere between 300-500 MT as against 300 MT of fossil fuel (90-91 period) – roughly of the same order.

For 1990, $CO_2$, and $CH_4$ inventories for this region amounts to about 200Tg C and 25Tg $CH_4$ respectively. $CO_2$-equivalent was around 3% of the world emissions. By 97-98, fuel consumption in India has already increased by 50%, amounting to about 800 Tg $CO_2$/yr for India alone. Sulfur content in Indian coal is low, so $SO_2$ emission is not as large as one would expect. SO2 emission for the region is about 5 $TgSO_2$/yr. Indian CO emissions are large, arising principally from large consumption of biofuels – large forest fires are rare – estimated to be around 60 Tg CO/yr. Black carbon and organic carbon emissions amount to 714 Gg and 1036 Gg respectively, a very large fraction of the global emissions.

**Pollution Scenario**

Several of the world's most polluted cities are found in South Asia: Calcutta, Delhi, Mumbai, Karachi, and Dhaka are examples of megacities that produce unacceptably high emissions of health endangering gaseous and particulate matter. A study of potential exposure to polluted outdoor air in Asia and the Pacific region showed that the percentage of population in cities exceeding WHO guidelines was as high as 98% for India and 99% for China.

Small particulate mater (SPM) values are several times higher than those prescribed by National Air Quality standards. For Delhi, annual average for 1997 was 370 $\mu gm^{-3}$, or 2 ½ times larger than the standard value for the residential area. Though the values of $SO_2$ and $NO_x$ generally remained within prescribed limits of 60-80 $\mu gm^{-3}$, there have been sharp increases in recent years – $SO_2$ by 109% from 1989 to 1996 and $NO_x$ by 82%. Vehicular pollution is a major contributing source – especially two-wheelers and diesel driven heavy trucks. For Delhi alone, vehicular population has increased more than tenfold in 20 years (from 0.2 M to 2.6 M) and account for more than 60% of the particulate pollution.

Haze is a common phenomenon in the region in wintertime even in high altitude areas: an example of haze seen during an aircraft flight in Nepal at a location close to Mt. There are evidences of sunlight reduction in several places with time: for Pune, there is a reduction by more than 12% during the period 1977-92. Although there is doubt about whether extreme climatic events have increased in frequency or in magnitude in recent years, one should note recent conditions. There have been two consecutive droughts in Pakistan and the northwestern parts of India while increased flooding situations in the high rainfall areas of Bangladesh, Nepal and the northeastern states of India. For Bangladesh, there have been severe floods at intervals of 7 to 10 years, the recent floods occurring in 1988 and 1998. During the 1998 flood, as much as 2/3 of the land area was inundated and nearly 1.6 M hectare of cropland was damaged.

These are serious concerns about the decreasing water availability in the region. Per capita water availability has decreased drastically – for Pakistan it is already less than 1700 $m^3$/cap i.e. it has reached water stress regime. A substantial part, 70-90% of available surface and groundwater, will continue to be required for agriculture, but there are ways to effect substantial saving in this sector.

There is distinct association between ambient air pollution and respiratory diseases, although estimates of mortality vary. Risk estimates for acute respiratory infections, chronic obstructive pulmonary disease and lung cancer, as well as tuberculosis, asthma are available. 500,000 premature deaths annually for children below 5 and adult women in India have been attributed to indoor pollution from biomass fuels. For south Asia, acute respiratory illness (ARI) deaths are reported to be as large as 1.4 million per year. Studies in India in a number of locations – Delhi, Calcutta, Mumbai, Lucknow also point to serious hazards from both indoor and outdoor pollution, although there are inherent uncertainties reflecting differences in pollution mixtures, sex, income level and occupational patterns.

## Indian Ocean Experiment (INDOEX)

### *Scientific Objectives for Studying Aerosols During INDOEX*

INDOEX was conducted over the tropical Indian Ocean from 1996 to 1999 during the winter monsoon seasons. The experiment culminated in an intensive field phase (IFP) in January-March 1999 during which an international group of over 200 scientists from the United States, Europe, India, and the Maldives using aircraft, ships, surface stations, satellites and computer models to study how air pollution affects the regional climate. INDOEX was designed to measure the characteristics, distribution, and radiative effects of these aerosols under present-day conditions. In addition to the IFP data, several years of chemical, meteorological, and radiative data are available from the (i) Kaashidhoo Climate Observatory (KCO, at 4.965° N, 73.466° E), (ii) annual cruises by Indian research vessels, and (iii) continental surface sites.

The Indian Ocean has been selected for measuring the transport, properties, and radiative effects of aerosols partly because of specific and favorable meteorological conditions. During the winter monsoon, the prevailing low-level winds in the northern Indian Ocean are northeasterly while the prevailing low-level winds in the southern Indian Ocean are southerly. These wind patterns transport continental and anthropogenic aerosols from India and Arabia over large areas of the Arabian Sea and northern Indian Ocean, and from India and Southeast Asia over large areas of the Bay of Bengal. The polluted air is advected as far south as the Inter-Tropical Convergence Zone (ITCZ). The convergence of pristine air from the southern Indian Ocean and polluted air from the northern Indian Ocean establishes a large gradient in aerosol loading close to the ITCZ. The expected geographic gradients in pollutants have been confirmed during pre-INDOEX ship experiments.

***Relationship of INDOEX of regional air pollution***

The radiative effects of aerosols over the Indian Ocean and Arabian Sea have been examined using observations from previous field programs. The first detailed calculations of the aerosol effects use data collected during the First Global Atmospheric Research program (GARP) Global Experiment (FGGE) and summer Monsoon Experiment (MONEX). These studies focus exclusively on the forcing by soil-dust aerosols from the Saudi Arabian peninsula and Sahara Desert. According to modern inventories of sulfate pollution from India, the flux of sulfur dioxide from India was approximately 1/3 the present-day value during MONEX. The results show that the dust aerosols enhance the shortwave heating rates by up to 0.3° C/day, a change of 50%, in the lower atmosphere over the Arabian Sea. The effects of anthropogenic aerosols obtained from INDOEX are substantially larger and will be discussed later. The climatic effects of aerosols released by the Kuwait oil fires during the 1991 Gulf War have also been extensively modeled, although of course the source of the aerosols was a transient phenomenon. More recent pre-INDOEX studies have examined the impacts of sulfate and carbonaceous aerosols on continental surface insolation. INDOEX, however, represents one of the first sustained international efforts to quantify the regional effects of anthropogenic aerosols from India, Southeast Asia, and China. The major funding for INDOEX was provided by the following agencies:

— U.S. National Science Foudation (lead agency)

— European Organisation for the Exploitation of Meteorological Satellites

— Indian Space Research Organisation, Bangalore

— Laboratoire de Météorologie Dynamique du Centre National de la Recherche Scientifique, Paris

— Max Planck Institute for Chemistry, Mainz
— U.S. Department of Energy
— U.S. National Aeronautics & Space Administration
— U.S. National Oceanic and Atmospheric Administration
— Vetlesen Foundation

## Air Pollution and Ozone

### Greenhouse Gases

The South Asian region has substantial populations living in rural locations, where domestic energy consumption depends on biofuels such as wood and cow dung, whereas in urban areas soft coke, kerosine and other liquid fuels are used as well. The economic development of the region, associated with an increasing demand for electricity, and the growing use of cars, cause substantial pollution emissions. At present, about one quarter of the energy use in Asia depends on biofuels while in India this fraction is even larger, close to 50%.

Until recently, North America and Europe dominated the use of fossil fuels, associated with strong carbon dioxide ($CO_2$) emissions and consequent global warming. The fossil fuel related $CO_2$ release per capita in Asia is nearly an order of magnitude smaller than in North America and Europe. However, Asian emissions are growing. Since about half the world's population lives in this region, the potential for strong pollution emissions is large. At present biomass burning is a main source of air pollution in the tropics. One objective of INDOEX was to improve the information on the mix of emission sources in South and Southeast Asia. It should be emphasized that estimating biomass burning emissions, notably from biofuel use, is difficult since they usually occur scattered over large rural areas. The burning process is not well defined because the fuel type and the combustion phase strongly affect the exhaust composition. It has been estimated that in India firewood contributes approximately two thirds to biofuel consumption, while the burning of dung and agricultural wastes contribute roughly equally to the remaining one third.

The types of compounds that are emitted by combustion processes are very much dependent on the temperature. At high burning temperatures much of the fuel is directly converted into $CO_2$. At lower temperatures more hydrocarbons and incompletely combusted gases, such as carbon monoxide (CO) and oxygenated hydrocarbons, are emitted. Domestic burning takes place at relatively low temperatures, and because it is a main source of pollution over India, CO concentrations are high in southern Asian air masses. In addition, INDOEX has shown that compounds that are typical for biomass burning are abundant. Examples of such products observed during INDOEX are acetone ($CH_3COCH_3$), methyl cyanide ($CH_3CN$) and methyl chloride ($CH_3Cl$). Especially methyl cyanide is a tracer of biomass burning emissions. Biomass burning is a strong source of air pollution as it is very inefficient as an energy source. The INDOEX measurements of CO have been reproduced through computer modeling.

Several models have subsequently been used to calculate CO concentrations for the entire region. Additional support for these model results is provided by a comparison with radiocarbon monoxide ($^{14}CO$) measurements performed from a ship during INDOEX. Fossil fuels are radiocarbon depleted, and the observed high $^{14}CO$ fraction indicates that 60-90% of the CO is of biogenic origin, attributed to the humaninduced burning of biomass.

Considering that the pollution occurs at low latitudes with high solar radiation intensity, one expects strong photochemical activity, possibly giving rise to ozone ($O_3$) buildup. Ozone is an important natural trace gas in the stratosphere (ozone layer) where it protects the Earth from harmful solar ultraviolet radiation. The lower atmosphere (troposphere) contains a much smaller fraction of $O_3$ (~10%), where it initiates the removal of pollutants through oxidation processes. The combination of CO, hydrocarbons and nitrogen oxides, notably from combustion processes, can lead to the formation of photochemical smog. "Photosmog" is typically associated with high levels of $O_3$, which (being a strong oxidant) can damage biological tissues, including the human respiratory tracts, agricultural crops and natural ecosystems. Ozone is also a greenhouse gas, so that its increase during industrialization contributes to climate change. Because of its important role in atmospheric chemistry, $O_3$ was measured from all platforms and ground stations, as well as through balloon soundings from the Kaashidhoo Climate Observatory (KCO) and two ships. Although $O_3$ concentrations near the Indian coast were about 50-70 nmol/mol and peak values even reached 80-100 nmol/mol, the air quality standards for $O_3$ were not strongly violated in polluted air from India.

The European Union air quality standard indicates that ozone may only exceed the threshold of 60 nmol/mol during a maximum of 25 days per year. Should values exceed 120 nmol/mol for more than an hour, the population must be warned. After three hours exceeding this value, government is to take countermeasures, for example, driving restrictions. Preliminary measurements in Indian cities indicate that $O_3$ rarely exceeds 80 nmol/mol, though there is some discrepancy between various measurements. This is consistent with the fact that much of the gaseous pollution originates from biomass burning.

In particular smoldering fires produce relatively little nitrogen oxides ($NO_x = NO + NO_2$), a necessary catalyst for photochemical $O_3$ formation. Nevertheless, substantial amounts of nitrate were measured in the aerosol particles (equivalent to several hundred pmol/ mol), and $NO_x$ levels of 5-15 nmol/mol have been measured in urban air, which indicates that $NO_x$ emissions are not negligible.

In spite of the fact that ozone in photosmog was not exceedingly high during INDOEX, it should be emphasized that only limited $O_3$ measurements are available in urban and none in suburban air, so that definitive statements cannot be made. Importantly, in the next decades emission trends in the region will likely reflect the increased use of fossil fuels, more strongly associated with $NO_x$ emissions, which will boost photochemical $O_3$ formation, possibly comparable to Europe and the USA during the past decades. Considering the population size, the situation in Asia will likely become even more serious. In general, the South Asian air quality problem will be largest during the winter monsoon in meteorologically stable and cloud free atmospheric conditions. The potential for future photosmog development during the dry monsoon is large. During the wet summer monsoon, however, under unstable and convective conditions, vertical mixing and transport of local pollution will limit the local violation of air quality standards, whereas much of the pollution will be introduced into the largescale circulation.

**Aerosols**

Part of the pollution emissions occur as microscopic particles (aerosols) or in particular as gaseous precursors of aerosols. In the latter case, chemical reactions within the atmosphere convert the primary gaseous pollution into gases with lower volatility some of which rapidly condense into particles. A main example is the release of sulfur dioxide ($SO_2$) from fossil fuel combustion. In the atmosphere, $SO_2$ is converted into particulate sulfate, which is largely removed by rain. A major discovery of INDOEX was the presence of a huge aerosol plume over the Indian Ocean, known as the INDOEX "brown cloud", sometimes reaching as far south as 10° S. The large extent of the plume is in part associated with the scarcity of precipitation during the winter monsoon. Evidently the large size of the plume must be related to very high aerosol concentrations near the sources. During INDOEX aerosol chemical and optical measurements were performed from the aircraft, ships and KCO. The latter is located on the Maldives about 500 km southwest of India and more than 1000 km from the main pollution centers. The analysis of filter samples collected at KCO shows a high average dry mass concentration of ~17 $\mu g/m^3$, while aircraft measurements indicate even higher concentrations.

The aerosol plume was typically composed of sulfate, organic compounds, black carbon (soot), mineral dust, ammonium, fly ash and minor compounds such as potassium and nitrate. Mass spectrometric particle analysis shows that the black carbon particles were always mixed with organics and sulfate. Very similar results were obtained from KCO, the aircraft and the ship measurements, which shows that the aerosol composition was remarkably uniform over the northern Indian Ocean.

The aerosol mass loading observed over the Indian Ocean, at least hundreds of kilometers downwind of the sources, is quite comparable to sub-urban air pollution in N-America and Europe. In N-America and Europe urban fine aerosols typically contain 28% sulfate, 31% organics, 9% BC, 8% ammonium, 6% nitrate and 18% other material (mean mass 32 $\mu g/m^3$), suburban aerosols 37% sulfate, 24% organics, 5% BC, 11% ammonium, 4% nitrate, 19% other material (mean mass 15 $\mu g/m^3$), and remote continental aerosols 22% sulfate, 11% organics, 3% BC, 7% ammonium, 3% nitrate, 56% other material (mean mass 4.8 μg/m³).

Moreover, the black carbon content was high. This gives the aerosol a strong sunlight absorbing character. The black carbon aerosol and fly ash are unquestionably human-produced since natural sources are negligible. Likewise, the sulfate can be largely attributed to anthropogenic sources. Filter samples collected on board a ship in the clean marine boundary layer over the southern Indian Ocean reveal a fine aerosol sulfate concentration of about 0.5 μg/m³, probably from the oxidation of naturally emitted dimethyl sulfide. The sulfate concentration over the northern Indian Ocean was close to 7 μg/m³, and we thus infer an anthropogenic fraction of more than 90%. Similarly, the ammonium concentration over the southern Indian Ocean, from natural ocean emissions, was 0.05 μg/m³, indicating an anthropogenic contribution of more than 95% to the nearly 2 μg/m³ of ammonium observed in the INDOEX pollution cloud.

It is more difficult to attribute the organic aerosol fraction to a particular source category. Secondary organic particles from natural hydrocarbon sources are probably of minor importance since India is scarcely forested. Moreover, the black carbon/total carbon ratio of 0.5, as derived from the filter samples, is typical for aerosols from fossil fuel combustion. In the aerosol over the clean southern Indian Ocean organic compounds were negligible, whereas over the northern Indian Ocean they amounted to almost 6 μg/m³. We thus infer that most of the particulate organics over the northern Indian Ocean were of anthropogenic origin. Free troposphere INDOEX aerosol components of natural origin included a total mass fraction of 1% sea salt and 10% mineral dust. Some of the mineral aerosol likely originated from road dust and agricultural emissions. Taken together, the human-produced contribution to the aerosol was at least 85%.

Similarly to gaseous air pollution, very little chemical information is available about (sub)urban aerosols from direct measurements in South and Southeast Asia. Preliminary observations of total aerosol mass in Calcutta indicate average concentrations of about 200 $\mu g/m^3$, whereas in Bombay average concentrations up to 2000 $\mu g/m^3$ are observed. Such high aerosol mass loadings strongly limit visibility, while air quality standards are exceeded by orders of magnitude. Air quality standards often refer to respirable suspended particulate matter (PM), being aerosols with a diameter smaller than 10 $\mu m$ ($PM_{10}$). The air quality standards for $PM_{10}$ typically range from 10-150 $\mu g/m^3$ annually averaged, depending on the particle composition and national legislation. Similarly to $O_3$, the potential for future increases is large, for example, through the growing use of sulfur containing fuels that also contribute to soot production. Further, it should be emphasized that the improvement of air quality measurements within regional networks, in conjunction with model calculations, addressing both local pollution in (sub)urban areas and downwind of large source areas (for example at KCO) are prerequisite to firmly establish the link between emissions and concentrations of gaseous and particulate air pollution.

**Spatial and Temporal Variations of the Haze**

The reduction of the direct (unscattered) solar radiation at the Earth's surface is related to the aerosol optical depth (AOD). The transmission of the direct solar radiation decreases exponentially with increasing optical depth. AOD is the best measure for column loading of aerosols. During INDOEX, the optical depth was obtained from surface and airborne radiometers, calculated from the chemical speciation using optical models, and derived from multi-spectral measurements combined with radiative transfer codes. The change in aerosol forcing with a unit increase in optical depth is called forcing sensitivity. The forcing sensitivity has been calculated both from INDOEX observations and models. One of the intrinsic properties of aerosols that determines atmospheric forcing is the fraction of sunlight incident on a typical aerosol particle which is scattered rather than absorbed. This property is the single-scattering albedo (SSA). In visible wavelengths SSA for sulfate aerosols is nearly identical to 1, while SSA for soot is approximately 0.3.

**Spatial Extent of Anthropogenic Aerosols**

The AOD has been measured in the southern and northern Indian Ocean during the winter months of 1996 to 1999 from the ORV *Sagar Kanya*. The optical depth in visible wavelengths varies from approximately 0.05 near 20° S to between 0.4 and 0.7 at 15° N. The value of 0.05 in the southern Indian Ocean is typical of unpolluted air, while values exceeding 0.2 are characteristic of polluted air. The large latitudinal gradient in optical depth is related to the variation in the number of particles per unit volume of air. The total number of particles changes from 250 $cm^{-3}$ in the southern Indian Ocean to roughly 1500 $cm^{-3}$ in the northern Indian Ocean. The two major natural aerosol species, sea-salt and dust, contribute only 11% and 12% (respectively) of the total optical depth in this region. As a result, approximately 80 (±10)% of the optical depth is obtained from anthropogenic aerosols. There is a transition between the polluted and undisturbed regions at approximately 5-8o S at the location of the ITCZ. Analysis of satellite observations for 1996 to 2000 indicates that the high optical depths are characteristic of the winter season during the late 1990s.

The optical depth averaged over the winter season for ocean regions between 0 to 30o N varies by less than 10% relative to the optical depth obtained during INDOEX in 1999.

The INDOEX observational estimates of optical depth are confined primarily to the Indian Ocean, Arabian Sea, and Bay of Bengal. Estimates of optical depth over land have been obtained from an aerosol assimilation model that is consistent with in situ measurements and satellite estimates of optical depth over oceans. The model is able to reproduce ship and surface measurements of daily-mean optical depth with an error of –0.03 to +0.07. During January through March of 1999, the model shows that the large optical depths over the northern Indian Ocean are accompanied by larger values over the adjacent continental areas where the anthropogenic aerosols originate. The time-mean optical depths over central and northern India, portions of Southeast Asia, and eastern China exceed 0.4 during this period. These values over land are the highest in the region bounded by 30° S to 30° N, 40° E to 120° E, which is consistent with the anthropogenic origin of the aerosols observed during INDOEX.

**Seasonal and Interannual Variability**

Variations of AODs during the winter monsoon (January to March) and summer monsoons (June to August) have been examined. Two major characteristics of the regional distribution can be identified from the maps. During the winter monsoon, the AODs are larger over both the Arabian sea and the Bay of Bengal. As identified in numerous studies the dominant sources for these aerosols are anthropogenic emissions. During the summer monsoon, when AODs are much larger over both regions the Arabian Sea and the Bay of Bengal. With the onset of the southwest monsoon, the Arabian plume becomes dominant and AODs off the Arabian and the north African coast reach values as high as 0.5 to 0.6, and spread to most of the Arabian sea. Three factors contribute to the summer time aerosol build up. First is the transport of mineral dust from Arabia and north Africa by the westerly winds at and above 700 mb. Second is the enhancement in the production of sea salt by the strong westerly winds. Lastly, the high humidity of the southwesterly surface winds increase the aerosol size which in turn leads to increase in scattering optical depth.

At ambient humidity, the single-scattering albedo (SSA) is approximately 0.86 to 0.9 for aerosols in the northern Indian Ocean. Several independent methods have been used to determine SSA from surface, ship-borne, and airborne instruments and from MACR calculations. The observed column averages span 0.86 to 0.9 and are remarkably consistent with the model estimates. These values are also consistent with column-average SSA obtained from surface stations at Goa and Dharwad in western India.

The values of SSA derived for INDOEX indicate that the aerosols in this region are a strong absorber of sunlight. Of the major aerosol species, only dust and black carbon are sufficiently absorptive to produce these low values of SSA. Black carbon is approximately 3.5 times as absorptive as dust, and the fine-particle concentration of black carbon exceeds that of dust by 40%. Therefore the relatively low values of SSA are explained primarily by the presence of black carbon.

Observations of the chemical composition of the aerosols confirm that the strong absorption is directly related to significant concentrations of black carbon introduced by fossil fuel and biomass burning. The natural sources of black carbon are negligible compared to the anthropogenic sources; and it is only 4% in the INDOEX region.

# 7

# Air Pollution from Ships

Today, ocean-going vessels transport 90 percent of all trade by volume to and from the 25 members of the European Community (EC), and nearly 80 percent by weight of all goods shipped in and out of the United States. Over the last three decades, activity in the marine shipping sector, as measured in metric ton-kilometers, has grown on average by 5 percent every year. Since emissions from ocean-going vessels have only been moderately controlled, this growth has been accompanied by a commensurate increase in the sector's contribution to local and global air pollution.

Ocean-going vessels contribute significantly to global emissions of nitrogen oxides ($NO_x$), sulfur oxides ($SO_x$), and particulate matter (PM). Indeed it is estimated that by 2020, ship emissions contributions to the European Union (EU) NOx and SOx inventories will surpass total emissions generated by all land-based mobile, stationary and other sources in the twenty-five nations. Air quality impacts from ocean-going vessels are especially significant in port cities and nations with extensive coastlines adjacent to shipping corridors. Studies making use of geographic marine activity data have estimated that about 70–80 percent of all ship emissions occur within 400 km (248 miles) of land.

Pollutants such as $NO_x$, $SO_x$, and PM have been linked to a variety of adverse public health outcomes, including increased risk of premature death from heart and pulmonary diseases and worsened respiratory disease. Marine emission sources are therefore responsible for a growing share of the public health impacts of exposure to air pollution in many regions. Although ocean-going vessels are among the most efficient modes of freight transport, they also generate substantial quantities of greenhouse gas emissions. Currently, carbon dioxide ($CO_2$) emissions from the international shipping sector as a whole exceed annual total greenhouse gas emissions from most of the nations listed in the Kyoto protocol as Annex I countries.

Relative to other sectors, the regulation of commercial marine vessels represents a significant political and legal challenge as ships operate largely outside of national boundaries. Ocean-going vessels are mainly subject to oversight by the International Maritime Organisation (IMO), under the purview of the United Nations. Unfortunately, IMO efforts to mitigate environmental impacts of emissions from global shipping have not kept pace with the industry's growth and the evolution of control technologies for controlling emissions.

The international process for establishing new regulatory requirements is further complicated by the complex relationships that exist between those nations to which most ships are registered under socalled "flags of convenience" and the large shipping interests that own most of the ships. As a result, the IMO adopted standards in 1997 that represented only a modest improvement in emissions from unregulated engines. When these standards entered into force they reflected levels already achieved by the average in-use engine. The IMO's current fuel sulfur limit of 4.5 percent is almost twice the average sulfur content of fuels in use in ships today and several thousand times the sulfur level of fuels used on-road in Europe and North America. These standards at best codify the industry's existing practices.

Under these circumstances, accelerated adoption of cleaner marine fuels and wider deployment of existing pollution control technologies and emission reduction strategies could dramatically improve the environmental performance of the shipping sector. To explore these opportunities, the ICCT undertook a review of the status of pollution control measures and programs implemented to date throughout the world.

Lower sulfur fuels, optimised engines, and exhaust after-treatment, such as selective catalytic reduction (SCR), have been shown to significantly improve the environmental performance of marine vessels. Other measures such as shore-side electricity and improved auxiliary engines can reduce so-called "hotelling" emissions—that is emissions generated while ships are docked at port. The feasibility and costeffectiveness of these measures has been demonstrated at several ports. Reducing marine NOx emissions are very cost-effective compared to remaining pollution control options for other mobile and stationary sources, especially in countries that have adopted a range of regulations to limit land-based emissions.

Nations in Europe and North America—along with port cities throughout the world—have deployed a suite of strategies to address air pollution from ships. These strategies have included regulations, voluntary programs, and marketbased programs. Examples of regulatory approaches have included national engine standards for the domestic vessel fleet and fuel sulfur standards for vessels operating in coastal waters and harbors. The voluntary harbor speed limits implemented in the San Pedro Bay by the ports of Los Angeles and Long Beach provide an example of a voluntary approach. Meanwhile, Sweden has experimented with a market-based approach by imposing a system of environmentally differentiated fairway and port dues that vary with ship emissions.

This successful program has led to increased use of lower-sulfur fuels and to the installation of SCR systems on a number of ships calling on Swedish ports. The recommendations advanced in this report identify implementation milestones in each of several distinct categories: (1) marine fuels, (2) new engines, (3) new vessels, (4) existing engines and vessels, (5) greenhouse gas emissions, (6) and in-port emissions. In the near-term, these recommendations generally call for widespread adoption of proven best available technologies in the 2010 timeframe. The ICCT's medium-term recommendations propose intermediary steps to be taken between 2012 and 2017.

Finally, technology-forcing, long-term recommendations are proposed for the post-2020 period. Implementing these recommendations will require the active engagement of numerous stakeholders, including ship owners and operators, ports, and regulators. Leadership from the businesses that demand shipping services is also crucial. Shipping customers are uniquely positioned to create incentives for improved performance in the shipping sector because they can require that their goods be transported with the least possible impact on the environment.

## Marine Fules

Reducing fuel sulfur content is an essential component of any strategy aimed at reducing SOx and PM emissions from marine vessels. Lower sulfur fuel also enables the use of advanced aftertreatment for NOx reductions. Existing plans to implement SOx Emission Control Areas

(SECAs), starting in 2006 in the Baltic Sea and expected in 2007 for the North Sea and English Channel, mean that a portion of the world's ships are now or will soon be using 1.5 percent sulfur fuels or equivalent after-treatment. In the short term, the ICCT recommends including other major shipping areas, such as the Mediterranean and parts of the North Atlantic and Pacific Rim, in the SECA program. Moreover, decisions concerning future SECAs should take into account sulfur- and particle-related public health impacts as well as impacts on land and sea ecosystems. Finally, ICCT recommends that the fuel sulfur limit in SECAs be lowered from 1.5 percent to 0.5 percent to achieve further emissions reduction in the 2010 timeframe and to facilitate the shift to lower sulfur fuels on a global scale. As a next step, the ICCT recommends that a uniform global fuel sulfur standard of 0.5 per-cent be introduced in the medium term. Relative to the 2.7 percent average sulfur content of current marine fuel, this step alone will reduce SOx emissions by approximately 80 percent and PM emissions by a minimum of 20 percent. At this level of fuel quality, selective catalytic reduction (SCR) will be fully enabled. Although SCR can function at higher fuel-sulfur levels, durability is significantly improved at lower levels.

Some uncertainty remains regarding the widespread availability of lower sulfur fuels in the recommended timeframe. However, there has been significant momentum among various stakeholders to reduce the global fuel sulfur limit. For example, some industry groups have recently expressed support for a global fuel standard requiring the use of 1 percent sulfur distillate fuel in the near term. In addition, current regulations in California and Europe require low-sulfur fuels in coastal waters, inland waterways, and at ports ahead of the ICCT-recommended dates. For example, the California auxiliary engine program requires the use of 0.5 percent sulfur fuel in the state's coastal waters and at port by 2007. The allowed sulfur level is lowered to 0.1 percent by 2010. Fuel with 0.1 percent sulfur content will also be required in ports and inland waterways in Europe by 2010 Adoption of a lower global fuel sulfur limit would provide the refining industry the clear signal it needs to invest in upgrading production facilities and ensure increased fuel availability. The ICCT also encourages further efforts to implement lower sulfur fuel ahead of the recommended schedule in coastal waters, inland waterways, and at ports. These programs can facilitate a transition to fleet-wide use of lower sulfur fuels while ensuring emissions reductions in proximity to the potentially impacted populations. In the long-term, fuel standards for marine fuels should be harmonised with standards for on-road fuels (500 ppm to 10–15 ppm).

## New Engines

The IMO's recent decision to review NOx standards for ocean-going vessels represents an opportunity to make significant progress in improving the performance of marine engines. The ICCT recommends requiring new engines to achieve NOx limits that are 40 percent lower than the current standard in the near term. This level can be reached primarily through engine upgrades. New engine standards should also be set to ensure significant reductions in PM emissions. A medium-term standard set at a level 95 percent below current standards for NOx would require the use of additional emission control technologies, including after-treatment controls. Further PM reduction should also be required. These near- and medium-term standards should be adopted at the same time to give manufacturers sufficient lead time to prepare for compliance and to direct their research and development activities accordingly. In addition to more stringent standards, the ICCT recommends that manufacturers be (1) required to certify engines using fuels that reflect actual in-use fuel quality; (2) be liable for in-use compliance and subject to in-use testing; and (3) be required to demonstrate the durability of emission control systems used to achieve compliance.

The production and use of engines that are significantly cleaner than the proposed standards should be encouraged both in the short and medium term through incentives to engine and technology manufacturers as well as vessel operators. Support for early technology demonstrations is necessary to ensure viable technology options are available to meet the increasingly stringent standards. In the long term, the ICCT recommends deploying incentives and other strategies to further promote the use of advanced technologies, especially technologies that achieve near-zero emissions, in promising applications.

### New Vessels

Many opportunities exist during a vessel's design and construction phases to make changes that would facilitate the use of low-emission control technologies. In the near term, the ICCT recommends that engine rooms be designed with enough space to allow for retrofit technologies including SCR as well as tank capacity for fuel switching in SECA and coastal areas. New vessels, especially ferries and cruise ships with regular routes and ports of call, should be built with the needed on-board equipment to utilise shore power when port-side facilities exist. Standardisation of international shore power requirements is also needed to ensure compatibility compatibility between shore-side facilities and ships. The ICCT supports the ongoing efforts within IMO to develop guidelines for shore-side electricity. In the long term, the ICCT encourages the use of advanced vessel design concepts that optimise energy efficiency as well as emissions performance and that incorporate propulsion from renewable energy sources including solar and wind power, where feasible.

**Vessels and Engines**

Control measures targeted at existing vessels and engines are necessary to significantly impact fleet-wide emissions. A low fleet turnover rate means that the largely uncontrolled vessels that make up the majority of the international marine shipping fleet today will continue to pollute for several decades before they are retired. Most existing control technology options have been developed and demonstrated on in-use vessels, suggesting that a large-scale retrofit program should be technically feasible. In the near term, the ICCT recommends that in-use standards reflecting best available control technologies be developed within the IMO.

These standards would allow, for example, future market-based programs (including the range of possible differentiated fee programs) to harmonise their emission requirements. The ICCT further recommends that any in-use standards used in market-based programs be designed to become more stringent over time so as to provide ongoing incentives for adopting the newest control technologies as they become available, proven, and cost-effective. The program should provide additional incentives to demonstrations of advanced technologies that provide emission reductions beyond the adopted in-use standards. Also in the short term, the ICCT recommends exploring the feasibility of early ship retirement as an extension of the ship recycling programs being developed by the IMO. If determined feasible, this type of program could be implemented in the medium to long term.

**Green House Gases**

The shipping sector's contribution to gases and particles that impact the Earth's climate is only beginning to be fully understood. Here, the ICCT recommends that near-term efforts focus on developing a baseline for the climate impacts of the world's vessel fleet. Once a baseline is established, market-based measures to reduce greenhouse gas emissions can be introduced, also in the near term. If cap and trade programs are developed for GHGs, they should only cover shipping sources and not include land-based sources. If the shipping sector becomes a source of credits for greenhouse gas emissions reductions, steps must be taken—as with any source of credits—to ensure that reductions are recognised only to the extent that they are quantifiable, enforceable, surplus to otherwise mandated reductions, and permanent. The ICCT also recommends that the IMO develop fuel economy standards for ships applicable to new vessels in the near term and existing vessels in the medium term.

## At Port

The ICCT recommends that emission mitigation measures should be adopted at all major port facilities and be fully integrated with local and/or regional air quality plans. Each port type has access to a range of implementation mechanisms to reduce emissions from ships at berth. For example, landlord ports can include emission reduction requirements in their lease agreements with tenant operators. Operating ports can directly implement some infrastructure measures.

Providing shore power is often the most effective emission-reduction option for vessels while at port. In some locations, however, pollution impacts from electricity generation may make this option less attractive. The ICCT recommends that port authorities and regulators select the strategy or combination of strategies that cost-effectively provides the most environmental benefits. If shore power does not meet these criteria, other options should be implemented including requiring hotelling ships to use the lowest sulfur on-road fuels available and/or engine emission controls. The implementation of shore power and alternative mitigation technologies should prioritise new terminals as well as those that are near residential areas.

In the medium-term, the ICCT recommends that incentives be provided for utilising lowcarbon sources for shore-side power (including renewable solar and wind generators). In the long-term, the development of cost-effective energy storage technologies and advanced low- or non-carbon generating options should make it possible to achieve near-zero hotelling emissions.

National and regional policy-makers are increasingly seeking to accelerate the introduction of emission control technologies and cleaner fuels into the international marine sector. Within the IMO process, several countries including Sweden, Norway, and Germany have emerged as proponents of further measures to reduce emissions from ships. The few environmental organisations that have obtained consultative status with the IMO have also been leading efforts to accelerate progress on these issues. Other environmental NGOs with related activities and expertise should consider applying for consultative status to bolster these efforts. Finally, these efforts within the IMO must be brought to the attention of the larger public. Greater public awareness of the environmental impacts of routine ship activity will undoubtedly result in added pressure to reduce emissions in much the same way that highly publicised oil spills led to an increased focus on accident prevention, impact mitigation, and accelerated phase-out of single-hull tanker ships by the IMO. Best practices and local or national successes should be shared with a global audience to demonstrate that dramatic reductions in emissions from marine vessels, both at sea and in port, are not only feasible but also cost-effective. In the end, collaboration between the public and private sectors and across a wide set of stakeholders will be essential to forge support for sustainable long-term measures to mitigate the public health and environmental impacts of shipping around the world.

# 8

# Toxic Molds and Threats to Indoor Air Quality

Molds are the most typical form of fungus found on earth, comprising approximately 25% of the earth's biomass.Other fungi include yeasts and mushrooms. Molds are ubiquitous on our planet and are essential decomposers of organic substances necessary for sustaining plant and animal life.Molds are made up of masses of filament-like cells called *hyphae*. Under the appropriate conditions, the *hyphae* will grow into long intertwining strings that form the main body of the fungus, or the *mycelium*. It is the mass of mycelium that is visible to the human eye. Molds reproduce via spores.

However, molds can also spread if a fragment of broken hyphae is transplanted to an area with adequate moisture and organic matter for food. What causes mold to grow? Molds are incredibly resilient and adaptable. Molds gain the nutrients they need through the decomposition of organic matter. Most molds found in indoor air are saprotropic, meaning they gather their food from dead moist organic matter such as wood, paper, paint, fabric, plant soil, dust, and cooked or raw foods.

However, molds can also grow on the surface of damp inorganic matter such as glass and bare concrete covered by an invisible *biofilm*. In short, all mold needs to germinate and grow is a readily available food source, undisturbed water, and time. Some molds can germinate in as little as four to 12 hours.Left undisturbed, mold can grow and spread in 24 to 72 hours.In their paper on health effects of molds, Peter Kozak, et al state, "Endogenous mold problems generally occur after prolonged or repeated water damage to a variety of organic materials." They further explain that mold spores can be brought into the home from outside. Outdoor factors that seem to affect indoor mold spore concentrations include marked shade, increased levels of available organic debris, and "natural or basically uncared for property."Floods, leaking pipes, leaking windows, and leaking roofs are all potential sources of moisture that can lead to mold infestation. Increased ambient humidity as a result of inadequate ventilation or improper drying of flooded areas can also lead to mold growth. Lifestyle choices such as overpopulating a residence, keeping a house closed up without running an air conditioner or dehumidifier, the presence of multiple indoor houseplants (especially if over watered and without adequate ventilation), and poor housecleaning habits can also lead to mold growth.

Molds reproduce through the production of spores. The environment in which a given mold may grow prolifically is very likely different from the environment necessary for spore production.After the spores are formed they are released into the air to be carried elsewhere for germination and growth. Mold spores can survive for many years in dry or hot environments, requiring only moisture and available organic matter to allow them to germinate.

### Toxic Molds

The term "toxic mold" when writing about molds that

have been implicated in severe health reactions in humans. As used in the press, this term generally refers only to those molds capable of producing *mycotoxins*. However, all molds under proper conditions are capable of eliciting a negative health response in humans through other methods such as inflammation, allergy, or infection.Therefore, limiting the discussion to just those molds that potentially produce mycotoxins is inappropriate.

**Properties of Molds**

Molds can elicit a variety of health responses in humans. The severity of the impact depends upon the type and amount of mold present as well as the susceptibility and sensitivity of the individual experiencing mold exposure.Humans are exposed to molds via ingestion, inhalation, and skin contact with mold or mold infested material.Although molds are living, multiplying organisms, they do not have to be alive to cause adverse health effects. Below is a list of mold components known to elicit a response in humans.

*Volatile organic compounds.* In the paper "Fungi & Indoor Air Quality" Sandra McNeel and R. Kreutzer state that "[m]olds produce a large number of volatile organic compounds. These chemicals are responsible for the musty odors produced by growing molds."Volatile organic compounds also provide the odor in cheese, and the "off" taste of mold infested foods. Exposure to high levels of volatile organic compounds, from any source, such as industrial work places, can irritate the mucous membranes and affect the central nervous system, producing such symptoms as headaches, attention deficit, inability to concentrate, and dizziness.

According to McNeel, at present the specific contribution of mold volatile organic compounds to building-related health problems has not been studied.

Also, mold volatile organic compounds are likely responsible for only a small fraction of total volatile organic compounds indoors.

*Allergens.* Due to the presence of allergens on spores, all molds studied to date have the potential to cause an allergic reaction in susceptible humans.Allergic reactions are believed to be the most common exposure reaction to molds. These reactions can range from mild, transitory responses, like runny eyes, runny nose, throat irritation, coughing, and sneezing; to severe, chronic illnesses such as sinusitis and asthma.

*Mycotoxins.* Some molds are capable of producing *mycotoxins,* natural organic compounds that are capable of initiating a toxic response in vertebrates.Molds known to potentially produce mycotoxins and which have been isolated in infestations causing adverse health effects include certain species of *Acremonium, Alternaria, Aspergillus,* Chaetomium, Cladosporium, Fusarium, Paecilomyces, Penicillium, Stachybotrys, and *Trichoderma.* This list is not all-inclusive.While a certain type of mold or mold strain type may have the genetic *potential* for producing mycotoxins, *specific environmental conditions* are believed to be needed for the mycotoxins to be produced. In other words, although a given mold might have the potential to produce mycotoxins, it will not produce them if the appropriate environmental conditions are not present. Currently, the specific conditions that cause mycotoxin production are not fully understood. The United States Environmental Protection Agency (U.S. EPA) recognises that mycotoxins have a tendency to concentrate in fungal spores and that there is limited information currently available regarding the processes involved in fungal spore release. As a result, the agency is currently conducting research on *Stachybotrys chartarum* in an effort to determine "the environmental conditions required for sporulation, emission, aerosolisation, dissemination and transport of *[Stachybotrys]* into the air."

**Effects of Molds**

Certain health effects, such as those related to allergic reactions like irritation of the eyes, nose, and throat, dermatitis, exacerbation of asthma, and respiratory distress, have been proven to be associated with mold exposure. Other reported effects such as fever, flu-like symptoms, fatigue, respiratory dysfunction, excessive and regular nose bleeds, dizziness, headaches, diarrhea, vomiting, liver damage, and impaired or altered immune function have been identified in persons who have been exposed to mold via inhalation, however, limitations in existing science hinder the ability of researchers to conclusively cite mold exposure as the cause of these health effects. Similarly, while kidney damage, infertility, reproductive cycle disruption, and neurotoxicity have been reported in animals exposed to molds under laboratory conditions, no evidence of these effects has been noted in humans.

The extent to which an individual may be affected depends upon his or her state of health, susceptibility to disease, the organism with which he or she came in contact, and the duration and severity of exposure. Some people experience temporary effects that disappear when they vacate infested areas.In others, the effects of exposure may be long-term or permanent.

Systemic infections caused by molds are not common.Normal, healthy individuals can resist systemic infection from airborne molds. Those at risk for systemic fungal infection are severely immunocompromised individuals such as those undergoing chemotherapy, individuals who have had organ or bone marrow transplants, and persons with HIV/AIDS.

A National Academy of Sciences report titled *Clearing the Air: Asthma and Indoor Air Exposures* determined there was sufficient evidence to show an association between mold exposure and the *aggravation* of asthma in individuals who are sensitised to the disease.

However, researchers report inadequate or insufficient evidence to determine whether or not an association exists regarding mold exposure and the *development* of asthma.An outbreak of Stachybotrys chartarum in Cleveland, Ohio, in 1994 was believed by some to have caused pulmonary hemorrhage in infants. Sixteen of the infants died. A Center for Disease Control sponsored a review of the cases concluded that the scientific evidence provided did not warrant the conclusion that inhaled mold was the cause of the illnesses in the infants. However, the panel also stated that further research was warranted, as the study design for the original research appeared to be faulty.

**Limitations of Molds**

Unfortunately, due to the variances in personal sensitivities and the vast array of molds it has been impossible to set exposure limits for molds that can be applied to all humans. Research regarding exposure limits and the effects of *ingested* molds is readily available. However, due to the limitations of currently available science, few studies exist regarding the exposure limits and effects of *inhaled* molds on individuals.This is due primarily to the absence of biomarkers specific to molds. Biomarkers are chemicals in the body which have a particular molecular feature that make it possible to measure if an individual has been exposed to disease causing organisms as well as the progress of disease or the effects of treatment. At present time there are no known biomarkers that can prove that an individual has been exposed to molds. Therefore, traditional means of proving the correlation between exposure and symptoms is impossible. The lack of biomarkers limits our ability to genuinely link indoor mold exposure to specific illnesses. New York City Department of Health Guidelines explain: "[S]usceptibility varies with the genetic predisposition, age, state of health, and concurrent exposures. For this reason, and because measurements of exposure are not standardised and biological markers of exposure are largely unknown, it is not possible to determine 'safe' or 'unsafe' levels of exposure for people in general."

Regarding infant exposure, the American Academy of Pediatrics Committee on Environmental Health in April 1998 stated that physicians should make every effort to ensure that infants less than one year of age avoid chronically moldy, water-damaged environments.

Several population-based studies make the correlation between mold exposure via inhalation and health effects. A comprehensive review of available literature relating to the health effects of mold performed by the Federal-Provincial Working Group on Mycology Air Quality in Public Buildings showed that such studies "have consistently detected an association between respiratory symptoms and home dampness and mould growth, but causality in these studies has not been established. Until the magnitude of population risk is known, it would be prudent, based on current evidence, to remediate indoor sources conducive to fungal growth."All resources reviewed agree that those most at risk for adverse health effects related to mold exposure are infants, the elderly, the chronically ill, severely immunocompromised individuals such as those undergoing chemotherapy, individuals who have had organ or bone marrow transplants, and persons with HIV/AIDS.

McNeel and Kreutzer explain that in the 1970s and 1980s mold contamination was identified as the primary cause for poor air quality in only 5% of more than 500

National Institute for Occupational Safety and Health (NIOSH) indoor air quality investigations. From 1986 to 1996, however, molds were the primary source of contamination 35-50% of the time. They note, "This change has been attributed at least partially to a paradigm shift from chemical contaminant-based investigations to an interdisciplinary approach combining evaluation of physical, chemical and microbial constituents of indoor air environments."In other words, until recently, we didn't look for molds.

According to Chris Gerber, an environmental consultant with nationally recognised testing and consulting firm Restoration Consultants, 70-80% of mold problems investigated by his employer in the past five to seven years have been in newly constructed buildings and newly constructed residences. Mr. Gerber attributes this high incidence of growth primarily to poor construction practices, the use of substandard materials, and increased air tightness of new construction techniques. He explains that increased air-tightness of newly constructed residences and office buildings can allow moisture to become trapped in exterior walls, creating an environment conducive to mold growth. Gerber further explains that in buildings with mold problems, centralised heating and air-conditioning systems can pick up contaminants and recirculate them throughout the building thus potentially spreading the infestation.

Jed Waldman, Chief of the California Department of Health Services Indoor Air Quality Program, suggests that regardless of construction practices, certain climate related events, such as flooding and the La Niña experienced two years ago, can cause even the most carefully constructed building to support mold growth.With regard to indoor air quality, molds have been implicated in office buildings, schools, and residences as contributing to the cause of illness in humans:

— *Sick Building Syndrome (SBS):* Although not attributed exclusively to molds, this is a term "used to describe situations in which building occupants experience acute health and comfort effects that appear to be linked to time spent in a building" because all other probable causes have been ruled out.Symptoms include headaches; eye, nose, and throat irritation; dry cough; dry or itchy skin; dizziness and nausea; difficulty in concentrating; fatigue; and sensitivity to odors.

  Sick building syndrome is attributed to inadequate ventilation, chemical contaminants from indoor and outdoor sources, and biological contaminants such as molds, bacteria, pollens, and viruses.

  In the paper, *Sick-Building Syndrome and Building Related Illnesses*, Theodore Passon, James W. Brown, and Seth Mante explain how increased air tightness of buildings in the 1970s, as a means of reducing energy consumption, has created environmental conditions conducive to the "proliferation of microorganisms [including mold] in indoor environments."Once growth has occurred, harmful organisms can be spread by improperly designed and maintained ventilation systems.

— *Poor Indoor Air Quality in Schools:* A 1998 survey of literature published on indoor air quality, ventilation and health symptoms in schools performed by Lawrence Berkeley National Laboratory states that health problems in schools related to indoor air quality are very similar to those symptoms reportedly experienced as sick building syndrome. The survey also revealed that microbiological pollutants, along with volatile organic compounds,

CO, and $CO_2$, were some of the most commonly measured air pollutants in schools, adding that fungal spores, bacteria, and allergens may be of particular concern. The survey cites water damage leading to mold contamination as the second most frequently reported building-related problem.

The United States government's General Accounting Office reports that nationwide one in five schools reports having indoor air quality problems. Consequences of poor indoor air quality in schools include: increasing risk of long and short-term health problems in teachers and students; a negative impact on students' ability to learn due to physical symptoms; reduced productivity of teachers; destruction of school equipment, including text books; and negative publicity for the school resulting in strained relationships among teachers, parents, and administrators.

— *Illness related to residential exposure:* Molds are believed to be one possible cause of certain severe adverse health impacts in children and adults in their homes. A few extreme examples include:

1. A Foresthill, California, couple, after realising remediation was too expensive, recently burned their home to the ground. The family claims that exposure to mold infestation caused respiratory problems, nose bleeds, rashes, and other illnesses in family members.They believe, in their case, that rebuilding is less costly than remediation.

2. Residents of four low-income housing apartment complexes in the Hunters Point section of San Francisco, California, blame toxic molds for a myriad of illnesses including rashes, hacking coughs, respiratory problems, headaches,

nosebleeds, and high blood pressure.Similar claims have occurred in other complexes statewide.

3. A Dripping Springs, Texas, couple's 11,500 square-foot home was quarantined after molds were alleged to have caused stomach problems, diarrhea, vomiting, severe respiratory scarring, and other maladies in their son, as well as severe memory loss in the husband, ultimately resulting in his inability to work.

While molds are commonly thought to grow only in warm, moist, dark environments, recent research has shown that mold can grow even in dry climates. Since 1991, the U.S. EPA has been conducting research to determine what environmental conditions permit mold infestation. The focus of the investigation has been on material properties, climate conditions, and microorganism interactions. According to the executive summary, "One of the most significant technical results from this project is that the effect of relative humidity is indirect and that very small amounts of moisture, well below those commonly cited, will permit growth."Therefore, while dry climates may experience a reduced incidence of mold growth, they are not immune to infestation.

## Prevention of Threat

There are no mandated actions specific to molds and indoor air quality required by any state or federal agencies. The U.S. EPA Indoor Air Quality website states, "Standards or Threshold Limit Values (TLVs) for airborne concentrations of mold, or mold spores, have not been set. Currently, there are no EPA regulations or standards for airborne mold contaminants."

*However, no mandates does not mean no action.* In response to public concern, some state and federal agencies have made available a variety of documents related to mold and indoor air quality. The majority of these publications focus on commercial building and school building infestation and remediation. However, there are a few publications available which are geared specifically toward the general public and residential exposure.

Every attempt was made to ensure that this list is all inclusive, however, it is possible that at the time of publication some documents may not have been discovered.

In response to increasing queries with regard to mold toxicity, the California Department of Health Services Indoor Air Quality Program has developed a website that includes a variety of documents related to this issue. Included in the selection is a document specific to residential exposure titled *Mold In My Home: What Do I Do?* The Occupational Health Branch of California Department of Health Services (OHB) is mandated to review new and emerging occupational hazards and propose new regulations to the California Division of Occupational Safety and Health (Cal OSHA). As a result of a proposal by the OHB, Cal OSHA is looking at adding molds to the sanitation standard for office buildings and workspaces.

The U.S. EPA has published recommendations on the cleanup of mold and control of moisture as part of its informational package titled "IAQ Tools for Schools." Federal guidelines regarding mold remediation are anticipated in a subsequent publication, "Mold Remediation in Schools and Commercial Buildings." EPA staff states that the guidelines are practical and thorough and focus on commercial buildings and schools. Staff also asserts that the guidelines have applicability to residential infestation as the underlying principles for remediation are the same. The document is anticipated to be available online at the U.S. EPA website by the third week in March 2001.

Several state and local agencies and universities across the United States have begun to address the issue of mold as an indoor air contaminant through the publication of articles and informational pamphlets discussing health effects, prevention, and remediation of indoor molds. Nationwide, agencies and universities offering informational materials include: Kansas Department of Health and Environment; Kansas State University; Minnesota Department of Health;

Montana State University; New Jersey Department of Health and Senior Services, Public Employees Occupational Safety and Health Program; New York City Department of Health, Bureau of Environmental & Occupational Disease Epidemiology; North Carolina State University; Texas Department of Health; and Washington State Department of Health.

The Canadian government has also conducted research regarding mold and indoor air quality. Proposition 65 requires that the Governor publish a list of chemicals that are known to the State of California to cause birth defects, reproductive harm, or cancer. Given that toxic molds are not classified as chemicals, they do not currently qualify for inclusion on the list.

### Testing and Remediation

Molds are resilient organisms. Because molds are capable of permeating porous materials surface cleaning is often insufficient to eradicate mold colonies. Also, because dead molds can damage human health, complete removal of all mold is required.

In a word, no. According to Harriett Ammann, a Senior Toxicologist for the Washington State Department of Health, "While testing for toxins is useful for establishing of disease, and adds to knowledge about mold toxicity in the indoor environment, prudent public health practice might advise speedy clean-up, or removal of heavily exposed populations from exposure as a first resort."

Further, according to John Banta, of Restoration Consultants, testing for mycotoxins that may be present in indoor air or structures is still experimental and not routinely offered.Banta and Gerber explain that most testing routinely performed at sites of infestation focuses on determining the type and quantity of mold present at the main site of mold infestation. This information is then used to determine to what extent the mold may have migrated throughout the building in an effort to reveal other, as yet not visible, locations of contamination.

Knowing the type of mold present or whether it produces toxins is not necessary to determine the appropriate method for clean up. Testing for type and quantity of molds is useful primarily when litigation is imminent (or in process), or in cases of severe health effects (for example, when physicians are trying to isolate an offending organism).

Unnecessary testing is also cost prohibitive. Determination of toxicity and differentiation of various mold strains requires expert testing. There is currently no home test for the layperson that will identify mycotoxin-producing molds. This is due primarily to the vast variety of molds that have multiple identity strains of which only some can potentially produce mycotoxins.According to Chris Gerber, of Restoration Consultants, a basic investigation for a single-family dwelling can start at $1,000 to $1,500. Testing in an attempt to document fault (for example, for use in determining financial responsibility in litigation) can increase the cost to $2,000 or more.

Molds are capable of eliciting a variety of adverse health effects. These effects are not limited only to exposure of mycotoxins, but are due to exposure of other fungal properties as well. Therefore, it is generally agreed that regardless of the type of mold present, any visible mold should be removed. Given the risks of large amounts of visible mold in cases of extreme growth, if testing is desired, remediation should not be postponed until all test results are complete unless the building is to be unoccupied and dehumidified in the interim. No State or Federal entities currently offer testing services of mold samples for mycotoxin production.

The New York City Department of Health was the first government entity to produce guidelines on mold. The document, "Guidelines on Assessment and Remediation of Fungi in Indoor Environments," has been cited by many as useful in planning mold cleanup procedures primarily because it was the only document available. However, some believe that the New York City guidelines are too arbitrary and focus too much on the size of visible mold when determining appropriate cleanup methods. For example, Jim Holland, of Remediation Consultants, states that the New York City guidelines do not take into consideration hidden molds which may be revealed when walls are opened. Thus, he claims, the guidelines can actually increase the risk of spreading the mold throughout the building as they do not require containment of small areas of infestation during the cleanup process.

Recommendations regarding mold remediation in commercial and school buildings are offered by a variety of federal, state, and private organisations and are not standardised. The soon-to-be-available U.S. EPA guidelines are widely anticipated and many hope they will clear the air with regard to differing schools of thought regarding mold cleanup in commercial and school buildings. However, there are certain aspects of mold cleanup which are agreed upon by all.

- Persons with respiratory problems, a compromised immune system, or fragile health, should not participate in cleanup operations.
- The source of moisture must be stopped and all areas infested with mold thoroughly cleaned. If thorough cleaning is not possible due to the nature of the material (porous versus semi- and non-porous), all contaminated areas should be removed.
- Cleanup crews should be properly attired. Mold should not be allowed to touch bare skin. Eyes and lungs should be protected from aerosol exposure.
- All efforts should be made to contain the infestation, while still allowing adequate ventilation for the cleanup crew during the cleanup process, in an effort to avoid spreading mold to other areas.

Experts consulted agree that a common sense approach should be taken when assessing mold growth. For example, it is generally believed that small amounts of growth, like those commonly found on shower walls, pose no immediate health risk to most individuals.

The California Department of Health Services information sheet, "Mold in My Home: What Do I Do?" contains guidelines for cleanup of large areas of mold that mirror those for commercial buildings and schools: persons with fragile health should not be involved in mold cleanup, wear protective gear, keep the area of infestation isolated, and maintain adequate ventilation for cleanup crew in area being cleaned. Residents are also cautioned to work over short time spans with rest periods in locations with access to fresh air, and to air out their home well after cleanup. This document also encourages residents to start the cleanup process with a small test patch. If the resident experiences any adverse health effects while cleaning the test area, it is advised that they consider hiring a professional to complete the job.

Some insurance companies will cover remediation if the precipitating event is considered insurable under the terms of the policy purchased by the property owner. Reading and understanding the coverage that is available under any given policy is advised. Because policy language can be difficult to interpret, in order to truly understand the coverage offered under their policy property owners should consult with an attorney specialising in insurance litigation. Precipitating events that may be considered insurable include water accumulation as a result of sudden and accidental flooding, wind-driven rain, fire damage, and sewer backup. Specific covered events are detailed in the homeowner's policy.

According to an article in the August 2000 issue of *Claims* magazine "Some of these losses are covered, some are not. If the water damage is the result of a covered loss, the resultant damage, mold (including fungi, mildew, etc), is probably also covered."In order to avoid paying for remediation of pre-existing mold this article urges adjusters to make a site visit within the first 24 to 72 hours, stating that any molds visible in the first two days post flooding are pre-existing and therefore not covered under the claim of loss.

Litigation to determine responsibility is not uncommon. Lawsuits have been filed against former property owners, builder's and insurance companies. Mealey's, a nationally recognised source for litigation information, released a publication in June 2000 which reports on 25 cases from around the country, with references to 29 available complaints, motions, and decisions in the field. These 25 cases represent only a small portion of what is believed, by one account, to be thousands of mold related cases nationwide.Litigation related to mold was the topic of a recent two-day conference in Philadelphia. Other similar conferences are planned across the country over the coming months.

**Prevention of Abnormal Growth**

All molds under proper conditions and concentrations are capable of adversely affecting human health. Therefore, it is clear, that reduction and prevention of mold exposure is needed to decrease the risk of damage to human health. Small amounts of mold are always present in the air, the key to decreasing adverse health affects from mold exposure is to prevent the germination and growth of mold on indoor surfaces. Individuals, landlords, building maintenance personnel, architects and builders need to know effective means of avoiding mold growth which might arise from lifestyle choices or maintenance and construction practices. Likewise, effective means of locating and remediating existing mold growths are paramount to decreasing the health affects of mold contamination.

All agree, the first step in preventing mold growth is to remove the source of moisture. Stopping leaks, decreasing indoor air humidity, and proper cleanup of waters after flooding are key to stopping the germination and growth of molds.In the meantime it is clear that once mold growth has occurred, fixing underlying leaks and immediate remediation of growth using proper techniques are required to avoid potential damage to health.

# 9

# Health Effects of Air Pollution

Air pollution causes ill health and death is well recognised. Air pollution is caused by both natural and man-made sources. Major man-made sources of ambient air pollution include industries, automobiles, and power generation. In indoor environments, tobacco smoke and combustion of solid fuels for cooking and heating are the most significant sources. In addition, construction material, furniture, carpeting, air conditioning, and home cleaning agents and insecticides can also be significant sources of chemical and biological pollutants indoors.

Fuel combustion is the primary source of a large number of health-damaging air pollutants, including fine and respirable particulate matter ($PM_{2.5}$ and $PM_{10}$), carbon monoxide (CO), sulfur dioxide ($SO_2$), nitrogen oxides ($NO_x$), volatile organic compounds (VOCs), ozone ($O_3$), and atmospheric lead. Some of these pollutants are direct by-products of fuel combustion, but others (such as $O_3$) are formed in the air through chemical reactions with other agents in the atmosphere.

Air pollution has both acute and chronic effects on human health. Health effects range anywhere from minor irritation of eyes and the upper respiratory system to chronic respiratory disease, heart disease, lung cancer, and death. Air pollution has been shown to cause acute respiratory infections in children and chronic bronchitis in adults. It has also been shown to worsen the condition of people with preexisting heart or lung disease. Among asthmatics, air pollution has been shown to aggravate the frequency and severity of attacks. Both short-term and long-term exposures have also been linked with premature mortality and reduced life expectancy.

Health impact of air pollution depends on the pollutant type, its concentration in the air, length of exposure, other pollutants in the air, and individual susceptibility. Different people are affected by air pollution in different ways. Poor people, undernourished people, very young and very old, and people with preexisting respiratory disease and other ill health, are more at risk. In cities, for instance, poor tend to live and work in most heavily polluted areas, and in rural areas poor are more likely to cook with dirtier fuels. In some countries, air quality standards tend to be more lax around industrial areas in cities, where many poor tend to live in squatter settlements. Poor also tend to be more malnourished, more likely to suffer from ill health and disease, and have less access to health care.

Air pollutants can also indirectly affect human health through acid rain, by polluting drinking water and entering the food chain, and through global warming and associated climate change and sea level rise. As a result of several decades of tighter emission standards and closer monitoring, levels of certain types of air pollutants have declined in many developed countries. Although, even at much reduced levels, air pollution continues to threaten public health in these countries. On the other hand, the ambient air pollution levels are a growing problem in urban centers in many developing countries. Several factors contribute to the worsening air pollution levels in developing-country cities, including rapid growth in urban population, increasing industrialisation, and rising demands for energy and motor vehicles. Other factors, such as poor environmental regulation, less efficient technology of production, congested roads, and age and poor maintenance of vehicles, also add to the problem.

In many rural areas of developing countries air pollution exposures tend to be highest indoors where majority of households use unprocessed solid fuels (biomass and coal) for cooking and space heating. These fuels are typically burned indoors in simple household cookstoves, such as a pit, three pieces of brick, or a U-shaped construction made from mud, which burn these fuels inefficiently and are often not vented with flues or hoods to take the pollutants to the outside. Under these conditions, high volumes of a number of health-damaging airborne pollutants are generated indoors, resulting in high exposures, especially among women who do the cooking and young children who stay indoors with mothers. The individual peak and mean exposures experienced in such settings are often much greater than the safe levels recommended by the World Health Organisation.

## Causal Mechanism

Causal biological mechanisms about health effects of air pollution are only partly understood. Air pollution exposure has been associated with compromised pulmonary immune defense mechanisms in both animals and humans. Tobacco smoking has also been shown to cause depressed immune system responses. Of the specific air pollutants, exposure to respirable particulate matter has been shown to induce a systemic inflammatory response involving stimulation of the bone marrow, which can contribute to cardiorespiratory morbidity. Other evidence indicates that exposure to polycyclic aromatic hydrocarbons—especially benzo[a]pyrene—can cause immune suppression and can increase the risk of infection and disease. Benzo[a]pyrene, a known carcinogen, also can increase the risk of lung and other types of cancers. Acute exposures to oxides of nitrogen and sulfur have been associated with increased bronchial reactivity and susceptibility to bacterial and viral infections. Carbon monoxide combines with hemoglobin to form carboxyhemoglobin, which reduces the oxygencarrying capacity of the blood and can contribute to anemia and adverse pregnancy outcomes, including miscarriage, stillbirth, low birth weight, and early infant mortality. In cases where biological pathways of effect are not known, questions remain as to how to establish causality.

## Approaches to Measuring Health Impacts of Air Pollution

*Prospective cohort studies.* The best way to measure the health impacts of air pollution in a given area is to conduct prospective epidemiologic studies by following a large cross-section of people over a long period of time. These studies should measure air pollution concentrations, individual exposure levels, ill health and death, as well as other factors, such as nutritional status, smoking habits, and family history of disease. Due to high costs and time involved, very few such studies have been carried out so far, and all such studies have been carried out in the developed countries. Two most commonly cited studies, the Harvard Six Cities study and the American Cancer Society study, both in the United States, have measured the effects of exposure to particulate matter on the survival function. These studies allow estimation of effect of air pollution on number of lifeyears lost.

*Time series studies.* Time series studies correlate daily variations in air pollution levels with variations in daily incidence of ill health, hospital admissions, or deaths in a given area. Time series studies measure the effects of acute air pollution exposure on morbidity and mortality, primarily among people more susceptible to adverse effects due to their poor nutritional status or due to a preexisting condition. Time series studies have the advantage that they do not need to control for many confounding factors, such as smoking, age structure, nutritional status, and access to health care, because these factors do not change much during the course of the study. Although in many developing countries, information on disease incidence, hospital admissions, and deaths itself may be correlated with such factors. Time series studies also have the advantage of being much less expensive than prospective cohort studies, but they fail to capture the chronic health effects of long-term exposure to air pollution. Also, time series studies tend to overestimate the effects on premature deaths because many of these people would have had short life expectancies in any case. Another problem with such studies is that they typically combine community-level air pollution data with individual-level health data and assume same exposure levels for all individuals in the community.

*Cross-sectional studies.* Retrospective/cross-sectional studies are used to compare health effects of air pollution across several locations at a given point in time (or in a given time period). These studies tend to be much cheaper and require less time than prospective cohort studies, and have the advantage over time series studies in that they capture both the acute and chronic effects of air pollution. A major constraint is that cross-sectional studies need to account for a large number of potentially confounding factors, such as differences in nutritional status, smoking rates, age distribution, and local industry.

*Meta-analyses.* In the absence of reliable epidemiologic studies of air pollution, meta-analyses of existing studies can be used to reduce uncertainty associated with individual studies and to obtain more reliable dose-response functions. Meta-analysis is a way to statistically pool results from several studies to obtain more reliable estimates of effects. For such pooling to be of good quality, studies need to be objectively selected to meet some basic standards of study design and data quality and to ensure that essential confounders have not been omitted. The recent comparative risk assessment effort of the World Health Organisation used this approach to estimate and rank burden of disease estimates for selected major risks to health, including the estimates of disease burden due to indoor and outdoor air pollution discussed below.

**Issues in Health Impacts**

The quantity and quality of scientific literature on the health effects of air pollution vary considerably by pollutant type and by health outcome. Following is a list of some of the issues in understanding health impacts of air pollution.

*Uncertainty.* According to the 2002 World Health Report, indoor air pollution from combustion of solid fuels for cooking and space heating is one of the eight most important risk factors in global burden of disease (measured as disability adjusted life years [DALY]). It accounts for an estimated 2.7% of the global disease burden and some 1.6 million premature deaths annually. In poor developing countries, indoor smoke from solid fuels ranks fourth, behind only under-nutrition, unsafe sex, and unsafe water/sanitation/hygiene, accounting for an estimated 3.7% of the disease burden. Urban air pollution additionally causes 1.4% of premature deaths and 0.8% of the global disease burden.

These estimates of disease burden are subject to considerable uncertainty, however. The estimates of disease burden due to urban air pollution are based only on the impact of air pollution on mortality. The impact of air pollution on morbidity could not be included in the calculations due to lack of reliable epidemiologic data. Similarly, disease burden estimates for indoor air pollution could not include several important health outcomes, such as tuberculosis, cataracts, asthma, and adverse pregnancy outcomes, that have been associated with indoor smoke but the evidence is not conclusive. These estimates also do not include health impacts of exposure to atmospheric lead and likely impacts of climate change.

*Transfer of results from one setting to another.* Most epidemiological studies of air pollution have been conducted in developed countries. Mainly due to the time and cost involved in establishing dose-response relationships from prospective epidemiologic studies, dose-response functions established in developed-country settings are often used in developing-country studies. How appropriate and applicable is this transfer of exposure-response functions? What are the options for developing countries that do not have good local epidemiologic studies of air pollution?

The level and composition of air pollutants vary considerably from one setting to the other, and overtime in the same setting. The relationships between ambient air pollution and health observed in developed countries at typically much lower levels of air pollution may not be applicable to developing-country situations with much higher air pollution levels. The disease composition in the underlying population may also differ between the settings. Moreover, the susceptibility of the populations to adverse health effects of air pollution may be quite different. Caution should be exercised when extrapolating the results from one study setting to another, keeping in mind the differences in pollutant mix, concentration levels, disease composition and prevalence levels, and susceptibility of underlying population.

A similar problem arises, when dose-response curves from ambient air pollution studies are applied to indoor air pollution studies in developing countries, where exposure levels tend to be many fold higher than from ambient sources. Also, the composition of air pollutants indoors may be very different from that outdoors. In some cases, exposure-response functions are derived from laboratory studies on animals, and questions remain about the uncertainty in applying these functions to humans.

Given that populations in developing countries are typically exposed to much higher air pollution levels and tend to have poorer nutritional status, higher prevalence of infectious diseases, and poorer access to health care, can we assume that the effects of air pollution are more severe for populations in developing countries? How do nutritional status and other factors modify the effects of air pollution on health?

*Valuing the health impacts of air pollution.* Another major issue relates to valuing the health impacts of air pollution. Policymakers often need to assess relative importance of air pollution in setting priorities for health programs. Estimating the costs of ill health, suffering, and premature death is a highly debated, controversial subject. The valuation of health impacts is usually done using 'willingness to pay' or 'cost of illness' approaches. Both these approaches need to make some heroic assumptions about value of lost life, or lost workdays, pain, and suffering. The task is made more difficult by non-availability of local epidemiologic studies, when additional assumptions about dose-response functions need to be made. Yet, policymakers cannot wait for more research, and decisions need to be made based on best available information, appropriately adapted to the local setting.

*Pollutant mixtures.* Given that air pollution is generated from many sources (including industry, motor vehicles, home cooking and heating, as well as from tobacco smoking and from numerous natural sources) and a variety of pollutants are present from each source, people are usually exposed to many air pollutants simultaneously. Air pollution contains complex mixtures of many pollutants, which vary overtime and from place to place depending on the sources of pollution and meteorological conditions. It is not clear if the effects of air pollution mixtures are additive or worse than the effects of individual pollutants. How do these effects differ in the short-term, and in the long term? How to disentangle the effects of various pollutants found together in the air? How to separate the short-term and long-term health effects of individual air pollutants and that of complex pollutant mixtures?

Much of the research on health impacts of air pollution has been conducted using single pollutants, and a majority of this research has focused on particulate matter. Is it appropriate to use one pollutant to represent a complex mixture of pollutants that people are exposed to? Does using several individual pollutants to estimate health effects in a study essentially leads to double counting and overestimation of effects?

*Measurement of exposure and ill health.* When estimating health risk from exposure to air pollution, it is critical to have good measures of exposure. In some studies, indirect surrogate measures of exposure are sometimes used. The examples include use of cooking fuel type as a measure of exposure to indoor smoke, traffic density for exposure to ambient air pollution, and chimney density as a proxy indicator for wood smoke pollution. How valid are these surrogate measures? Also, there are problems in assessing individual exposures from area samplers and community monitoring stations. Accuracy of exposure measurement equipment also raises questions about validity and reliability of results and their comparability across studies.

Measurement of health parameters also poses many challenges for the study of health impacts of air pollution. For example, in many developing countries, vital statistics and cause of death data are not routinely collected or are not reliable. Clinical data are also highly unreliable because there is often considerable selection due to differential access to health services and awareness of health problems. Multiple sources of health care, varying quality of services, poor diagnosis, and poor record keeping further complicate the study of health impacts.

*Thresholds.* Other questions relate to the shape of the dose-response curves. It is not entirely clear if the relationships between different air pollutants and health outcomes are linear? Are there thresholds below which no adverse health effects are expected to occur? For which pollutants and health outcomes?

*Politics.* Health effects of air pollution are studied more in developed countries, and more for more visible ambient air pollution. Health effects of indoor air pollution are also being investigated mostly for developed countries for indoor bio-allergens, such as dust mites and molds. However, the indoor air pollution problem is many orders of magnitude greater in poor, rural homes of developing countries, where half of the world population relies on unprocessed solid fuels for cooking and heating and is exposed to very high levels of many air pollutants on a daily basis. Because indoor air pollution from cooking and heating is often worst in poor, remote, rural areas, it tends to receive less attention than more visible ambient air pollution in cities and other proximate health problems that catch public attention. This raises questions about objectivity in the selection of topics to study health impacts of air pollution and availability of funding for such research.

## Future Directions

Needless to say that the science of air pollution and health linkages is less than conclusive. The biological mechanisms through which air pollutants affect health are not fully understood, and there is limited epidemiologic research on air pollution, limited mostly to the developed countries with very different pollution levels, exposure patterns, and underlying population characteristics than in developing countries. While there has been considerable research on health impacts of tobacco smoking, and there is growing evidence on the effects of ambient air pollution, the research on health effects of indoor air pollution from household use of unprocessed solid fuels is particularly weak.

Clearly, there is need to strengthen both the quantity and quality of evidence linking air pollution and various health outcomes, especially for developing countries and for health conditions with weak or no evidence. This can be accomplished by measuring exposure levels more directly; by including clinical measures of disease outcomes; and by adequately accounting for social, behavioral, nutritional, and environmental confounding factors. In these efforts, there is need to use more powerful study designs, such as prospective cohort studies and randomised intervention trials, designed specifically for areas with weak data. A randomised intervention trial currently underway in upland Guatemala to study effects of improved stoves on acute respiratory infections in young children is a good example.

In addition to looking at health effects of individual pollutants, epidemiologic studies need to examine the effects of different pollutant mixtures and interactions between components of pollutant mixtures. It is also important to identify groups that are more susceptible to adverse health effects of air pollution, and to understand the interactions with poverty, nutritional status, and other factors.

In the meantime, policymakers in many developing countries need to design programs, set standards, and take action to mitigate adverse health effects of air pollution. These efforts need to carefully adapt available knowledge from other settings, keeping in mind the differences in pollutant mixtures, concentration levels, exposure patterns, and various underlying population characteristics.

Air pollution is a serious and growing problem in many developing countries, particularly in rapidly expanding cities, where pollution levels are often several times the maximum level recommended by the World Health Organisation. That high levels of air pollution are bad for health and can cause premature death is well known. But policymakers need more specific information to set priorities. Should available funds go to air pollution reduction—or would they be better spent on sanitation, education, AIDS prevention, or any of a host of other pressing concerns? A cost-benefit analysis of air pollution reduction measures is a necessary input to such a decision. However, answers vary greatly depending on the method used. Some of the pitfalls that commonly arise when attempting to estimate the dollar value of benefits associated with a given reduction in pollution levels in low-income countries. Because cost-benefit estimates are often necessary despite these pitfalls, boxes describe how to prepare a rough estimate of the benefits of pollution reduction, using Delhi, India as an example.

**Measuring the Effects of Air Pollution**

Air pollutants affect health in a variety of ways, from itchy eyes to premature death. Although air pollution affects both death and illness rates, greater social costs are likely to be associated with premature death. Chronic exposure to air pollution can lead to premature death by increasing the rate at which lung tissue ages, by contributing to chronic obstructive lung disease, and by exacerbating cardiovascular disease. Short-term peaks in air pollution can increase the chance that a person in a weakened state or a susceptible person will die. The most accurate way to measure the impact of air pollution on death rates in a given area is to conduct epidemiologic studies for that area. To capture both acute and chronic effects, a large cross-section of individuals should be followed prospectively for at least ten years, measuring both the air pollution concentrations to which they are exposed and other factors that affect the risk of death (smoking behavior, body mass index, family history).

Because such studies are expensive, only two have been completed to date, both in the United States. These studies show how pollution alters the survival function and allow the number of life-years lost as a result of air pollution to be calculated.

Because they are less expensive, retrospective studies that use aggregate data are common. Cross-sectional studies correlate aggregate variations in air pollution levels across counties with death rates by county. Cross-sectional studies are generally unreliable, however, for two reasons. First, if there has been recent immigration, historic air pollution levels in a county may not measure the concentrations to which people currently living in the county were exposed. Second, it is usually hard to control for confounding factors (smoking behavior, dietary risk factors) at the county level.

Time series studies, which correlate daily variation in air pollution with variation in daily mortality in a given city, avoid the problems mentioned above, but at best measure the effects of acute exposure to air pollution on death rates. Because they measure primarily the effect of acute exposure to air pollution on people who are in a weakened state and are therefore susceptible to death as a result of a sudden rise in pollution concentrations, such studies measure only some of the effects of air pollution on premature death. To the extent that many of these people would have had short life expectancies even in the absence of pollution, time series studies may overestimate the life-years lost as a result of air pollution. Such studies can provide a lower-bound estimate of the reduction in premature death that would accompany a reduction in air pollution, however.

In the United States half of all deaths are caused by cardiovascular disease or respiratory illness. In Delhi, India, by contrast, fewer than 20 percent of all deaths are attributable to these causes. Even if people in Delhi reacted to a given change in particulate pollution in exactly the same way that people in the United States did, the effect of the change on total mortality would be lower in Delhi.

**Risk of Dealth**

Economists calculate the value of a reduction in risk of death by determining how much an individual is willing to pay for it. One approach to measuring willingness to pay is to infer the value from compensating wage differentials in the labor market. The theory behind this approach is simple: other things being equal, workers in riskier jobs are compensated with higher wages for bearing more risk. A second approach to measuring willingness to pay is to ask people what they would pay to reduce their risk of dying. This is referred to as the contingent valuation approach.

Recent compensating wage studies in the United States have produced estimates of the value of a statistical life in the range of \$1.9–\$10.7 million. Contingent valuation studies have produced slightly lower estimates of \$1.2–\$9.7 million. Compensating wage studies measure compensation for risk of instantaneous death for people who are about 40 years old and thus measure the approximate value of 35 years of life. Because death from air pollution reduces life-years by less than 35 years on average, labor market estimates should be adjusted accordingly.

Compensating wage and contingent valuation studies have not yet been conducted in developing countries. Premature death in these countries is typically valued by using estimates of the value of a statistical life from the United States and adjusting for income differences between the two countries, or by calculating forgone earnings. Which method is preferable? Economic theory tells us that the willingnessto-pay approach yields a value of a statistical life that exceeds the value of forgone earnings as long as people are risk averse. In the United States, for example, estimates of the value of a statistical life are typically five to ten times higher than the value of forgone earnings. If people in other countries were equally risk averse, then it would be appropriate to multiply the value of forgone earnings in the country of inter-est by the same factor found in U.S. studies. It is, however, likely that risk aversion varies with living standards and that the value of a statistical life in developing countries is a smaller multiple of forgone earnings than it is in the United States. Until further research provides estimates of the value of a statistical life in developing countries, one approach to valuation is to use forgone earnings as a lower bound to the value of changes in life expectancy and projections of the value of life from U.S. studies as an upper bound, as shown in box 2.

**Air Pollution with Particulate Matter, Ozone and Nitrogen Dioxide**

In most countries in Europe, ambient air quality has improved considerably in the last few decades. However, there is a large body of evidence suggesting that exposure to air pollution, even at the levels commonly achieved nowadays in European countries, leads to adverse health effects. In particular, exposure to pollutants such as particulate matter and ozone has been found to be associated with increases in hospital admissions for cardiovascular and respiratory disease and mortality in many cities in Europe and other continents. Recent studies have also tried to quantify the health effects caused by ambient air pollution; e.g., within the "Global Burden of Disease" project of the World Health Organisation (WHO) it has been estimated that worldwide, close to 6.4 million years of healthy life are lost due to long-term exposure to ambient particulate matter.

In the 1990s, WHO updated its Air quality guidelines (AQG) for Europe, to provide detailed information on the adverse effects of exposure to different air pollutants on human health. The prime aim of these guidelines was to provide a basis for protecting human health from effects of air pollution. The guidelines were in particular intended to provide information and guidance for authorities to make risk management decisions. The European Union (EU) used the WHO guidelines as a basis to set binding air quality limit values and target values for all EU member states for several pollutants.

Since the most recent update of the WHO AQGs, there have been many new studies published that have investigated the effects of air pollution on human health. In order to provide (European) policy makers with state-of-the-art knowledge on the effects of air pollution on human health, it was considered necessary to review the new evidence systematically. At this stage, the review concentrated on the following pollutants: particulate matter (PM), ozone ($O_3$) and nitrogen dioxide ($NO_2$). In particular, the question under discussion was whether there was sufficient new evidence to reconsider the current WHO guidelines.

***Particulate Matter (PM)***

Airborne particulate matter represents a complex mixture of organic and inorganic substances. Mass and composition in urban environments tend to be divided into two principal groups: coarse particles and fine particles. The barrier between these two fractions of particles usually lies between 1 μm and 2.5 μm. However, the limit between coarse and fine particles is sometimes fixed by convention at 2.5 μm in aerodynamic diameter ($PM_{2.5}$) for measurement purposes. The smaller particles contain the secondarily formed aerosols (gas-to-particle conversion), combustion particles and recondensed organic and metal vapours. The larger particles usually contain earth crust materials and fugitive dust from roads and industries. The fine fraction contains most of the acidity (hydrogen ion) and mutagenic activity of particulate matter, although in fog some coarse acid droplets are also present. Whereas most of the mass is usually in the fine mode (particles between 100 nm and 2.5 μm), the largest number of particles is found in the very small sizes, less than 100 nm. As anticipated from the relationship of particle volume with mass, these so-called ultrafine particles often contribute only a few % to the mass, at the same time contributing to over 90% of the numbers.

Particulate air pollution is a mixture of solid, liquid or solid and liquid particles suspended in the air. These suspended particles vary in size, composition and origin. It is convenient to classify particles by their aerodynamic properties because: (a) these properties govern the transport and removal of particles from the air; (b) they also govern their deposition within the respiratory system and (c) they are associated with the chemical composition and sources of particles. These properties are conveniently summarised by the aerodynamic diameter, that is the size of a unit-density sphere with the same aerodynamic characteristics. Particles are sampled and described on the basis of their aerodynamic diameter, usually called simply the particle size.

The size of suspended particles in the atmosphere varies over four orders of magnitude, from a few nanometres to tens of micrometres. The largest particles, called the coarse fraction (or mode), are mechanically produced by the break-up of larger solid particles. These particles can include wind-blown dust from agricultural processes, uncovered soil, unpaved roads or mining operations. Traffic produces road dust and air turbulence that can stir up road dust. Near coasts, evaporation of sea spray can produce large particles. Pollen grains, mould spores, and plant and insect parts are all in this larger size range.

The amount of energy required to break these particles into smaller sizes increases as the size decreases, which effectively establishes a lower limit for the production of these coarse particles of approximately 1 μm. Smaller particles, called the fine fraction or mode, are largely formed from gases. The smallest particles, less than 0.1 μm, are formed by nucleation, that is, condensation of low-vapour-pressure substances formed by high-temperature vaporisation or by chemical reactions in the atmosphere to form new particles (nuclei). Four major classes of sources with equilibrium pressures low enough to form nuclei mode particles can yield particulate matter: heavy metals, elemental carbon, organic carbon and sulfates and nitrates. Particles in this nucleation range or mode grow by coagulation, that is, the combination of two or more particles to form a larger particle, or by condensation, that is, condensation of gas or vapour molecules on the surface of existing particles. Coagulation is most efficient for large numbers of particles, and condensation is most efficient for large surface areas. Therefore the efficiency of both coagulation and condensation decreases as particle size increases, which effectively produces an upper limit such that particles do not grow by these processes beyond approximately 1 μm. Thus particles tend to "accumulate" between 0.1 and 1 μm, the so-called accumulation range.

Sub micrometre-sized particles can be produced by the condensation of metals or organic compounds that are vaporised in high-temperature combustion processes. They can also be produced by condensation of gases that have been converted in atmospheric reactions to low-vapour-pressure substances. For example, sulphur dioxide is oxidised in the atmosphere to form sulphuric acid ($H_2SO_4$), which can be neutralised by $NH_3$ to form ammonium sulfate. Nitrogen dioxide ($NO_2$) is oxidised to nitric acid ($HNO_3$), which in turn can react with ammonia ($NH_3$) to form ammonium nitrate ($NH_4NO_3$). The particles produced by the intermediate reactions of gases in the atmosphere are called *secondary particles*. Secondary sulphate and nitrate particles are usually the dominant component of fine particles. Combustion of fossil fuels such as coal, oil and petrol can produce coarse particles from the release of non-combustible materials, i.e. fly ash, fine particles from the condensation of materials vaporised during combustion, and secondary particles through the atmospheric reactions of sulphur oxides and nitrogen oxides initially released as gases.

Recently a comprehensive report on PM phenomology in Europe was compiled. Sulfate and organic matter are the two main contributors to the annual average $PM_{10}$ and $PM_{2.5}$ mass concentrations, except at kerbside sites where mineral dust (including trace elements) is also a main contributor to $PM_{10}$. On days when $PM_{10} > 50$ μg/m$^3$, nitrate becomes also a main contributors to $PM_{10}$ and $PM_{2.5}$. Black carbon contributes 5–10% to $PM_{2.5}$ and somewhat less to $PM_{10}$ at all sites, including the natural background sites. Its contribution increases to 15–20% at some of the kerbside sites.

Because of its complexity and the importance of particle size in determining exposure and human dose, numerous terms are used to describe particulate matter. Some are derived from and defined by sampling and/or analytic methods, e.g. "suspended particulate matter", "total suspended particulates", "black smoke". Others refer more to the site of deposition in the respiratory tract, e.g. "inhalable particles", which pass into the upper airways (nose and mouth), and "thoracic particles", which deposit within the lower respiratory tract, and "respirable particles", which penetrate to the gas-exchange region of the lungs. Other terms, such as "PM10", have both physiological and sampling connotations.

#### *Ozone ($O_3$)*

Ozone is the most important photochemical oxidant in the troposphere. It is formed by photochemical reactions in the presence of precursor pollutants such as $NO_x$ and volatile organic compounds. In the vicinity of strong $NO_x$ emission sources, where there is an abundance of NO, $O_3$ is "scavenged" and as a result its concentrations are often low in busy urban centres and higher in suburban and adjacent rural areas. On the other hand, $O_3$ is also subject to long-range atmospheric transport and is therefore considered as a trans-boundary problem. As a result of its photochemical origin, $O_3$ displays strong seasonal and diurnal patterns, with higher concentrations in summer and in the afternoon. The correlation of $O_3$ with other pollutants varies by season and location.

There is evidence from controlled human and animal exposure studies of the potential for $O_3$ to cause adverse health effects. Epidemiological studies have also addressed the effects of short and long-term exposures to $O_3$ and provided important results. However, the health effects of $O_3$ have been less studied than those of PM and thus more research is needed, especially addressing the spatial and seasonal patterns and misclassification of individual exposure in association with health outcomes.

# 10

# Air Pollution and Climate Change

Urban air pollution has a significant impact on the chemistry of the atmosphere and thus potentially on regional and global climate. Already, air pollution is a major issue in an increasing number of megacities around the world, and new policies to address urban air pollution are likely to be enacted in many developing countries irrespective of the participation of these countries in any explicit future climate policies. The emissions of gases and microscopic particles (aerosols) that are important in air pollution and climate are often highly correlated due to shared generating processes.

Most important among these processes is combustion of fossil fuels and biomass which produces carbon dioxide ($CO_2$), carbon monoxide (CO), nitrogen oxides ($NO_x$), volatile organic compounds (VOCs), black carbon (BC) aerosols, and sulfur oxides ($SO_x$, comprised of some sulfate aerosols, but mostly $SO_2$ gas which subsequently forms white sulfate aerosols). In addition, the atmospheric lifecycles of common air pollutants such as CO, $NO_x$ and VOCs, and of the climatically important methane ($CH_4$) and sulfate aerosols, both involve the fast photochemistry of the hydroxyl free radical (OH). Hydroxyl radicals are the dominant "cleansing" chemical in the atmosphere, annually removing about 3.7 gigatons (1 gigaton = $10^{15}$ gm) of reactive trace gases from the atmosphere; this amount is similar to the total mass of carbon removed annually from the atmosphere by the land and ocean combined.

## A Chemistry Primer

The ability of the lower atmosphere (troposphere) to remove most air pollutants depends on complex chemistry driven by the relatively small amount of the sun's ultraviolet light that penetrates through the upper atmospheric (stratospheric) ozone layer. This chemistry is also driven by emissions of $NO_x$, CO, $CH_4$ and VOCs and leads to the production of $O_3$ and OH. Figure 1 reviews, with much simplification, the chemical reactions involved. The importance of this chemistry to climate change occurs because it involves both climate-forcing greenhouse gases ($H_2O$, $CH_4$, $O_3$) and air pollutants (CO, NO, $NO_2$). It also involves aerosols ($H_2SO_4$, $HNO_3$, BC) that influence climate (through reflecting or absorbing sunlight), productivity of ecosystems (through their exposure to $O_3$, and to $H_2SO_4$ and $HNO_3$ in acid rain), and human health (through inhalation). Also important are free radicals and atoms in two forms: very reactive species like $O(^1D)$ and OH, and less reactive ones like $HO_2$, $O(^3P)$, NO and $NO_2$.

Referring to Figure 1, when OH reacts with $CH_4$ the CH4 is converted mostly to CO in steps that consume OH and also produce $HO_2$. The OH in turn converts CO to $CO_2$, $NO_2$ to $HNO_3$, and $SO_2$ to $H_2SO_4$. The primary OH production pathway occurs when $H_2O$ reacts with the $O(^1D)$ atoms that come from dissociation of $O_3$ by ultraviolet (UV) light. Within about a second of its formation, on average, OH reacts with other gases, either by donating its O atom (*e.g.*, to CO to form $CO_2$ and H) or by removing H (*e.g.*, from $CH_4$ to form $CH_3$ and H2O). The H and $CH_3$ formed in these ways attach rapidly to $O_2$ to form hydroperoxy ($HO_2$) or methylperoxy ($CH_3O_2$) free radicals which are relatively unreactive. If there is no way to rapidly recycle $HO_2$ back to OH, then levels of OH are kept relatively low. The addition of $NO_x$ emissions into the mix significantly changes the chemistry.

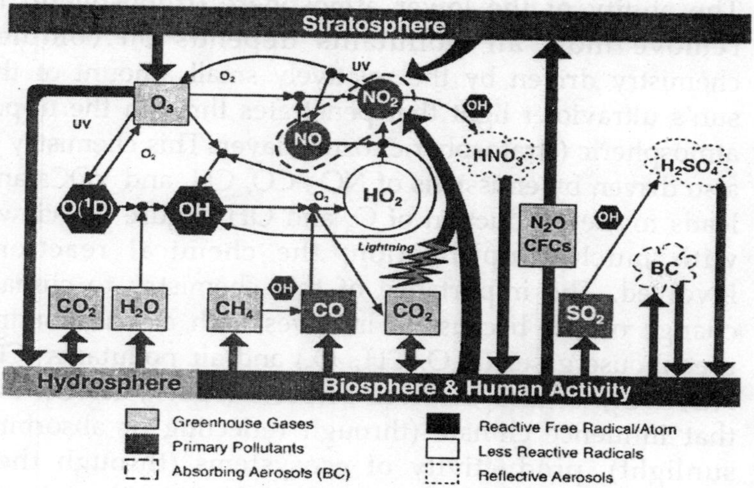


*Figure 1*

Specifically, a second pathway is created in which NO reacts with $HO_2$ to form $NO_2$ and to reform OH. Ultraviolet light then decomposes $NO_2$ to produce O atoms (which attach to $O_2$ to form $O_3$) and reform NO. Hence $NO_x$ (the sum of NO and $NO_2$) is a catalyst which is not consumed in these reactions. The production rate of OH by this secondary path in polluted air is about five times faster than the above primary pathway involving $O(^1D)$ and $H_2O$. The reaction of NO with $HO_2$ does not act as a sink for $HO_x$ (the sum of OH and $HO_2$) but instead determines the ratio of OH to $HO_2$. Calculations for polluted air suggest that $HO_2$ concentrations are about 40 times greater than OH. This is due mainly to the much greater reactivity of OH compared to $HO_2$.

If emissions of air pollutants that react with OH, such as CO, VOCs, $CH_4$, and $SO_2$, are increasing, then keeping all else constant, OH levels should decrease. This would increase the lifetime and hence concentrations of $CH_4$. However, increasing $NO_x$ emissions should increase tropospheric $O_3$ (and hence the primary source of OH), as well as increase the recycling rate of $HO_2$ to OH. This OH increase should lower $CH_4$ concentrations. Thus changing the level of OH causes greenhouse gas, and thus climate, changes. Climate change will also influence OH. Higher ocean temperatures should increase $H_2O$ in the lower troposphere and thus increase OH production through its primary pathway. Higher atmospheric temperatures also increase the rate of reaction of OH with $CH_4$, decreasing the concentrations of both. Greater cloud cover will reflect more solar ultraviolet light, thus decreasing OH, and vice versa.

Added to these interactions involving gases, are those involving aerosols. For example, increasing $SO_2$ emissions and/or OH concentrations should lead to greater concentrations of sulfate aerosols which are a cooling influence. Accounting for all of these interactions, and other related ones, requires that a detailed interactive atmospheric chemistry and climate model be used to assess the effects of air pollution reductions on climate.

## Integrated Global System Model

For our calculations, we utilise the MIT Integrated Global System Model (IGSM). The IGSM consists of a set of coupled submodels of economic development and its associated emissions, natural biogeochemical cycles, climate, air pollution, and natural ecosystems. It is specifically designed to address key questions in the natural and social sciences that are amenable to quantitative analysis and are relevant to environmental policy. The current structure of the IGSM is shown in Figure 2.

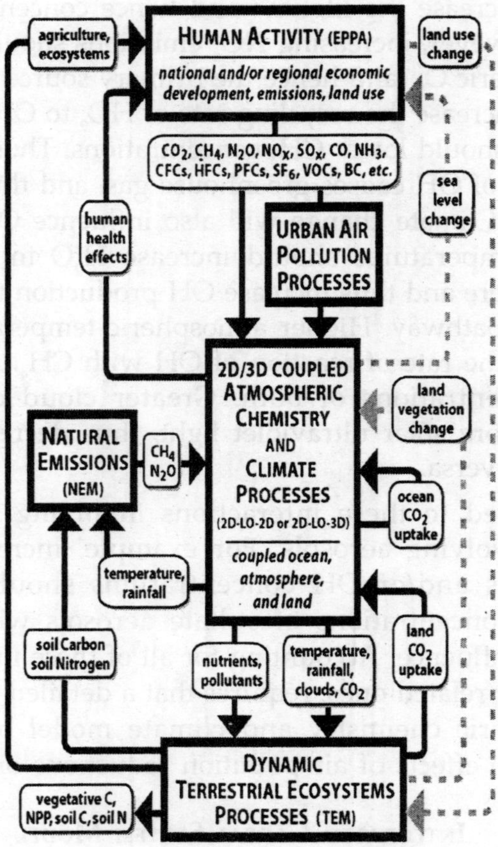


*Figure 2. Schematic illustrating the framework, submodels, and processes in the MIT Integrated Global System Model (IGSM)*

Chemically and radiatively important trace gases and aerosols are emitted as a result of human activity. The Emissions Prediction and Policy Analysis (EPPA) submodel incorporates the major relevant demographic, economic, trade, and technical issues involved in these emissions at the national and global levels. Natural emissions of these gases are also important and are computed in the Natural Emissions Model (NEM) which is driven by IGSM predictions of climate and ecosystem states around the world.

The coupled atmospheric chemistry and climate submodel is in turn driven by the combination of these anthropogenic and natural emissions. This submodel includes atmospheric and oceanic chemistry and circulation, and land hydrological processes. The atmospheric chemistry component has sufficient detail to include its sensitivity to climate and different mixes of emissions, and to address the effects on climate of policies proposed for control of air pollution and vice versa. Of particular importance to the calculations presented here, the urban air pollution (UAP) submodel is based upon, and designed to simulate, the detailed chemical and dynamical processes in current 3D urban air chemistry models. For this purpose, the emissions calculated in the EPPA submodel are divided into two parts: urban emissions which are processed by the UAP submodel before entering the global chemistry/climate submodel, and non-urban emissions which are input directly into the large-scale model.

The UAP enables simultaneous consideration of control policies applied to local air pollution and global climate. It also provides the capability to assess the effects of air pollution on ecosystems, and to predict levels of irritants important to human health in the growing number of megacities around the world. The atmospheric and oceanic circulation components in the IGSM are simplified compared to the most complex models available, but they capture the major processes and, with appropriate parameter choices, can mimic quite well the zonal-average behavior of the complex models. We use the version of the IGSM with 2D atmospheric and 2D oceanic submodels here, although the latest version has a 3D ocean to capture better the deep ocean circulations that serve as heat and $CO_2$ sinks. The 2D/2D version we use here resolves separately the land and ocean (LO) processes at each latitude and so is referred to as the 2D-LO-2D version.

The outputs from the coupled atmospheric chemistry and climate model then drive a Terrestrial Ecosystems Model which calculates key vegetation properties including production of vegetation mass, land-atmosphere $CO_2$ exchanges, and soil nutrient contents in 18 globally distributed ecosystems. TEM then feeds back its computed $CO_2$. fluxes to the climate/atmospheric chemistry submodel, and its soil nutrient contents to NEM, to complete the IGSM interactions. The current IGSM does not include treatment of black carbon (BC) aerosols. Detailed studies with a global 3D chemistry and climate model indicate multiple, regionally variable and partially-offsetting, effects of BC on absorption and reflection of sunlight, reflectivity of clouds, and the strength of lower tropospheric convection. These detailed studies also suggest important BC-induced changes in the geographic pattern of precipitation, not surprisingly since aerosols have important and complex effects on cloud formation, and on whether clouds will even produce precipitation. Methods to capture these effects in the IGSM are currently being explored. In light of the difficulty in simulating these and other regional effects, the numerical results presented here are limited to temperature and sea level effects, primarily at the global and hemispheric level.

## Numerical Experiments

To investigate, at least qualitatively, some of the important potential impacts of controls of air pollutants on temperature, we have carried out runs of the IGSM in which individual pollutant emissions, or combinations of these emissions, are held constant from 2005 to 2100. These are compared to a reference run in which there is no explicit policy to reduce greenhouse gas emissions. Specifically, in five runs of the IGSM, we consider caps at 2005 levels of emissions of the following air pollutants:

(1) $NO_x$ only

(2) CO plus VOCs only

(3) $SO_x$ only

(4) Cases (1) and (2) combined

(5) Cases (1), (2) and (3) combined

Cases (1) and (2) are designed to show the individual effects of controls on $NO_x$ and reactive carbon gases (CO, VOC), although such individual actions are very unlikely. Case (3) addresses further controls on emissions of sulfur oxides from combustion of fossil fuels and biomass, and from industrial processes. Cases (4) and (5) address combinations more likely to be representative of a real comprehensive air pollution control approach.

One important caveat in interpreting our results is that we are neglecting the effects of air pollutant controls on: (a) the overall demand for fossil fuels (*e.g.*, leading to greater efficiencies in energy usage and/or greater demand for non-fossil energy sources), and (b), the relative mix of fossil fuels used in the energy sector (*i.e.* coal versus oil versus gas). Consideration of these effects, which may be very important, will require calculation in the EPPA model of the impacts of $NO_x$, CO, VOC and $SO_x$ emission reductions on the cost of using coal, oil, and gas. Such calculations have not yet been included in the current global economic models (including EPPA) used to address the climate issue. Such inclusion requires relating results from existing very detailed studies of costs of meeting near-term air pollution control to the more aggregated structure, and longer time horizon, of models used to examine climate policy.

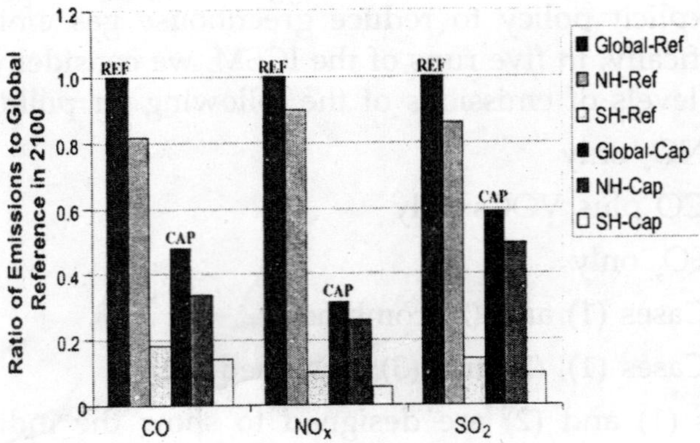


*Figure 3*

In Figure 3 we show the ratios of the emissions of $NO_x$, CO/VOC, and $SO_x$ in the year 2100 to the reference case in 2100 when their emissions are capped at 2005 levels. Because these chemicals are short-lived, the effects of their emissions are largely restricted to the hemispheres in which they are emitted. Figure 3 therefore shows hemispheric as well as global emission ratios. For calibration, the reference global emissions of $NO_x$, CO/VOC, and $SO_x$ in 2100 are about 5, 2.5, and 1.5 times their 2000 levels.

**Effects on Concentrations**

In Figure 4, the global and hemispheric average lower tropospheric concentrations of $CH_4$, $O_3$, sulfate aerosols, and OH in each of the above five capping cases are shown as percentage changes from the relevant global or hemispheric reference. From Figure 4a, the major global effects of capping $SO_x$ are to decrease sulfate aerosols and slightly increase OH. Capping of $NO_x$ leads to decreases in $O_3$ and OH and an increase in $CH_4$. The CO and VOC cap increases OH and thus increases sulfate and decreases $CH_4$. Note that CO and VOC changes have opposing effects on $O_3$ so the net changes when they are capped together are small. Combining $NO_x$, CO and VOC caps leads to an $O_3$ decrease and a slight increase in CH4. Finally, capping all emissions causes substantial lowering of sulfate aerosols and $O_3$ and a small increase in $CH_4$.

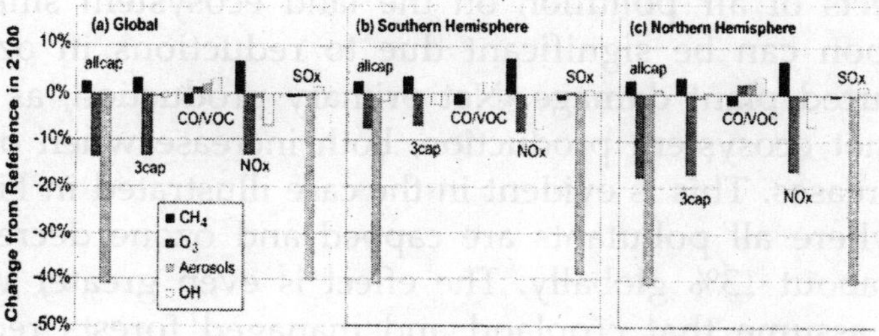


*Figure 4*

The two hemispheres generally respond somewhat differently to these caps due to the short air pollutant lifetimes and dominance of northern over southern hemispheric emissions. The northern hemisphere contributes the most to the global averages and therefore responds similarly. The southern hemisphere shows very similar decreases in sulfate aerosol from its reference when compared to the northern hemisphere when either $SO_x$ or all emissions are capped.

When compared to the southern hemisphere, the northern hemispheric ozone levels decrease by much larger percentages below their northern hemisphere reference when either $NO_x$, $NO_x$/CO/VOC, or all emissions are capped. Capping $NO_x$ emissions leads to significant decreases in OH and thus increases in methane in both hemispheres. Because methane has a long lifetime relative to the interhemispheric mixing time, its global concentrations are influenced by OH changes in either hemisphere alone, or in both. Hence $CH_4$ also increases in both hemispheres when $NO_x$/CO/VOC or all emissions are capped even though the OH decreases only occur in the northern hemisphere in these two cases.

### Effects on Ecosystems

Effects of air pollution on the land ecosystem sink for carbon can be significant due to reductions in ozone-induced plant damage. Net primary production, as well as net ecosystem production, both increase when ozone decreases. This is evident in the case illustrated in Figure 5 where all pollutants are capped and ozone decreased by about 13% globally. The effect is even greater when we assume that cropland and managed forests receive optimal levels of nitrogen fertiliser. The land sink (NEP) is increased by 30 to 49% or 0.6 to 0.9 gigatons of carbon (in $CO_2$) in 2100 through the illustrated pollution caps. These ecosystem calculations do not include the additional positive effects on NPP and NEP of decreased acid deposition and decreased exposure to $SO_2$ and $NO_2$ gas, that would result from the pollution caps considered. They also do not include the negative effects on NPP and NEP of decreasing nutrient nitrate and possibly sulfate deposition that also arise from these caps.

### Economic Effects

If we could confidently value damages associated with climate change, we could estimate the avoided damages in dollar terms resulting from reductions in temperature due to the lowered level of atmospheric $CO_2$ caused by the above increases in the land carbon sink achieved with the ozone caps. We could similarly value the temperature changes due to caps in other pollutants besides ozone.

However monetary damage estimates suffer from numerous shortcomings. Felzer *et al.* valued increases in carbon storage in ecosystems due to decreased ozone exposure in terms of the avoided costs of fossil fuel $CO_2$ reductions needed to achieve an atmospheric stabilisation target. The particular target they examined was 550 ppm $CO_2$.

The above extra annual carbon uptake of 0.6 to 0.9 gigatons of carbon is only 2 to 4% of year 2100 reference projections of anthropogenic fossil $CO_2$ emissions. However, as these authors point out, this small level of additional uptake can have a surprisingly large effect on the cost of achieving a climate policy goal. Here we conduct a similar analysis using a 5% discount rate, and adopting the policy costs associated with 550 ppm $CO_2$ stabilisation, to estimate the policy cost savings that would result from the increased carbon uptake through 2100 in the "allcap" compared to the "ref" scenarios shown in Figure 5. The savings are \$2.5 to \$4.7 trillion. These implied savings are 12 to 22% of the total cost of a 550ppm stabilisation policy.

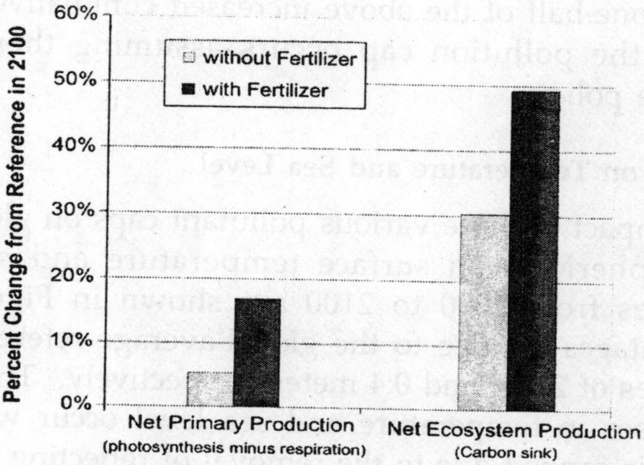


*Figure 5. Net annual uptake of carbon by vegetation alone (net primary production) and vegetation plus soils (net ecosystem production, the land carbon sink) for the $NO_x/SO_x/CO$ plus VOC capped (allcap) case is shown for the year 2100 as a percentage change from the reference case.*

The disproportionately large economic value of the additional carbon uptake has two reasons. One reason is that the fossil carbon reduction savings are cumulative; the total reference 2000-2100 carbon uptake is 36 (without Fertiliser) and 75 (with Fertiliser) gigatons, or about 6 to 13 years of fossil carbon emissions at current annual rates. A second reason is that the additional uptake avoids the highest marginal cost options. This assumes that the implemented policies would be cost effective in the sense that the least costly carbon reduction options would be used first, and more costly options would only be used later if needed. An important caveat here is that, as shown in Felzer *et al*, a carbon emissions reduction policy also reduces ozone precursors so that an additional cap on these precursors associated with air pollution policy results in a smaller additional reduction, and less avoided ecosystem damage. A pollution cap as examined here, assuming there was also a 550ppm carbon policy in place, leads to only a 0.1 to 0.8 gigaton increase in the land sink in 2100 and a cumulative 2000-2100 increase of carbon uptake of 13 to 40 gigatons of carbon, which is about one-half of the above increased cumulative uptake when the pollution cap occurs assuming there is no climate policy.

### Effects on Temperature and Sea Level

The impact of these various pollutant caps on global and hemispheric mean surface temperature and sea level changes from 2000 to 2100 are shown in Figure 6 as percentages relative to the global-average reference case changes of 2.7°C and 0.4 meters respectively. The largest increases in temperature and sea level occur when $SO_x$ alone is capped due to the removal of reflecting (cooling) sulfate aerosols. Because most $SO_x$ emissions are in the northern hemisphere, the temperature increases are greatest there. For the $NO_x$ caps, temperature increases in the southern hemisphere, but decreases in the northern hemisphere. For CO and VOC reductions, there are small decreases in temperature driven by the accompanying aerosol increases and $CH_4$ reductions, with the greatest effects being in the northern hemisphere where most of the CO and VOC emissions occur. When $NO_x$, CO, and VOCs are all capped, the nonlinearity in the system is evidenced by the fact that the combined effects are not simple sums of the effects from the individual caps. Ozone decreases and aerosol increases (offset only slightly by $CH_4$ increases) lead to even less warming and sea level rise than obtained by adding the CO/VOC and $NO_x$ capping cases. Finally the capping of all emissions yields temperature and sea level rises that are smaller but qualitatively similar to the case where only $SO_x$ is capped, but the rises are greater than expected from simple addition of the $SO_x$-capped and CO/VOC/$NO_x$-capped cases. Nevertheless, the capping of CO, VOC and $NO_x$ serves to reduce the warming induced by the capping of $SO_x$.

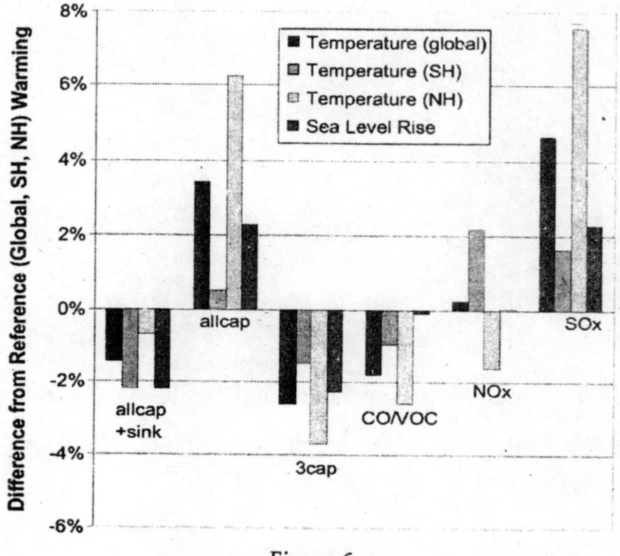


*Figure 6*

Note that these climate calculations in Figure 6 omit the cooling effects of the $CO_2$ reductions caused by the lessening of the inhibition of the land sink by ozone. This omission is valid if we presume that anthropogenic $CO_2$ emissions, otherwise restricted by a climate policy, are allowed to increase to compensate for these reductions. To illustrate the lowering of climate impacts if we allowed the sink-related $CO_2$ reductions to occur, we show a sixth case in Figure 6 ("allcap+sink") which combines the capping of all air pollutant emissions with the enhanced carbon sink from Figure 5. Now we see that the sign of the warming and sea level rise seen in the "allcap" case is reversed in the "allcap+sink" case.

# 11

# Impact of Air Pollution on Forest Ecosystems

The International Union of Forest Research Organisations (IUFRO) 22nd biennial meeting for Specialists in Air Pollution Effects on Forest Ecosystems "Forests under Anthropogenic Pressure – Effects of Air Pollution, Climate Change and Urban Development" was held September 10–16, 2006, in Riverside, CA. The meeting was hosted by the USDA Forest Service, Pacific Southwest Research Station, on behalf of IUFRO Research Group 7.01.00 "Impacts of Air Pollution and Climate Change on Forest Ecosystems". A satellite symposium on "Forest Fires and Air Pollution Issues" was also held.

## Forest Fires

The recent increase in application of fire as a tool for land-use change has resulted in more frequent occurrences of extended fire and smoke episodes worldwide with serious consequences for human and ecosystem health. Some of these events have been associated with droughts that may be attributed to increasing regional climate variability. International efforts are needed to address the underlying causes and avoidance of excessive fire application, and to implement sound fire and smoke management practices.

Fires, both wild and prescribed, can significantly contribute to ambient air pollution. Consequently, air quality concerns and regulations are major constraints to the implementation of prescribed fire strategies in many instances. Modeling is an important tool for characterising the transport of polluted air masses resulting from forest fires, and for predicting forest fire effects on air quality. Comparisons between air pollutant concentrations estimated with and without forest fires may allow us to assess forest fire contribution to ambient air pollution. Consequently, there is an urgent need to develop predictive models to assist land and air resource managers in the timely and controlled use of fires that would not cause exceedances of criteria air pollutant standards.

Fire has emerged as a major issue in forests with Mediterranean-type climates where incidence, duration, and severity of fires have significantly increased. Climate warming may be contributing to this increase. For example, in the U.S., the combination of 90 years of intensive fire suppression and significantly warmer air temperatures since the 1990s has led to successive years of record fire occurrence. Although the ecological changes in forests caused by wildfire are profound and generally well known, their effects on ecosystems may be more pronounced and long lasting in regions with warming climate. The potential for prescribed fire to mitigate nitrogen (N) saturation resulting from atmospheric deposition is of particular interest. There are only few other disturbances that can cause comparable N loss in such a short amount of time. The possibility for prescribed fire to mitigate Nsaturated sites needs to be explored with field experiments. However, there is also a clear need to understand the fate of released N gases and aerosols, and their potential effects on air quality and N deposition downwind from the fires. Furthermore, fire can change base cations from organic compounds to oxides and hydroxides in ash that, in turn, become readily available for replenishment of base cations on exchange sites, increasing soil short-term acid buffering capacity. Increases in soil pH are commonly observed following fire.

The increase in forest susceptibility to fires caused by air pollution is also of interest. Elevated background levels of $O_3$ and N deposition increase leaf turnover rates, as well as N and lignin content in needles. This reduces litter decomposition and increases fuel buildup. Both $O_3$ exposure and elevated available N may reduce root/shoot mass and plant control of water loss, thus increasing tree susceptibility to drought stress and fire. Better understanding of mechanisms of air pollution effects on wild-land fires in forests and other terrestrial ecosystems is needed.

The sources of forest fire emissions and effects of forest fires on carbon budget should be quantified. Improved understanding of the effects of changing fire regimes on all aspects of the carbon cycle is needed before we can better predict the magnitude, or even the direction, of the effect of changing fire regimes on the global carbon budget.

## Climate Change

Research on air pollution effects on forests has often been focused on the impacts of air pollutants on trees. However, forest ecosystem health encompasses much greater complexity than simply tree health. The potential for air pollution effects on biogenic VOC emissions, phyllosphere biota, interorganismic interactions, and soil fauna must also be considered when assessing forest health. The impacts of air pollution and climate change on forest ecosystems have been traditionally studied and reported on separately. Many air pollutants and greenhouse gases ($O_3$ is both) have common sources, contribute to global warming through increased radiative forcing, interact in the atmosphere, and jointly affect ecosystems. Their combined effects may significantly differ from a sum of separate effects. Some pollutants that are now of emerging concern may be of significant importance for forest health. These include, in particular, secondary organic aerosols, especially aerosol deposition rates to forests and mechanisms for increasing leaf wetness.

The technical facilities to test air pollution effects on plant systems have been developed: greenhouses, climate chambers, open-top chambers, dynamic leaf cuvettes, branch and single tree chambers, EDU application, free-air fumigation systems, and sapflow measurements. Sapflow can be a useful tool to validate stomatal functions and to assess whole-plant O3 uptake, even though time offsets may exist between sapflow and meteorological parameters. There has been a significant effort to move away from chamber studies to field studies in forest health indicator development, which has improved our understanding of the scope and extent of air pollution effects in forested ecosystems. Together with innovative technical facilities for pollutant exposure or measurement, new tools for data analysis and presentation including dynamic general vegetation models, meta-analysis, and redundancy canonical analysis are now available.

Presentations were dedicated to elevated carbon dioxide ($CO_2$), N deposition and availability, climatic factors, pathogens, acidification, heavy metals, and especially to $O_3$. More than a third of the 80 oral and poster presentations were on $O_3$ effects, which remains the phytotoxic pollutant of most concern.

There was much discussion about $O_3$ critical levels in an attempt to improve regulatory standards. Both North Americans and Europeans use exposure-based metrics, while the Europeans are moving toward the use of flux-based metrics for $O_3$, with the latter gaining momentum. The definition of critical levels has changed over the years to allow for the incorporation of the flux concept. A key component of flux-based critical levels is knowledge about the receptor including growth, yield, visible injury, and seed production. One of the most pressing challenges in setting an $O_3$ standard involves our ability to make $O_3$ critical levels more mechanistic by quantifying both stomatal $O_3$ flux and the plant's defense capacity. Flux alone is not sufficient, because plant response depends on nonlinear mechanisms, such as defense, ontogeny, and phenology.

Practical application of a flux-based risk assessment requires estimates of these parameters that are species specific. However, data for parameterisation of flux-based critical level models are missing for most plant species. An interesting contribution was presented for the Mediterranean forest trees. Results from leaf-level or canopy-level measurements showed significant deviations from the currently used flux model, which has implications for $O_3$ flux calculations. These results demand adequate validation of models before they are used to predict productivity losses. Free-air $O_3$ release systems at field sites can provide information on underlying mechanisms.

To understand the integrated effects of air pollution and climate change on forest ecosystems, a multifactorial and cause/effect-related approach is required. Experiments that focused on more than one stress factor simultaneously (e.g., reporting on complex responses in cases of interactions between air pollution, climate, and forest pests) resulted in novel results that will have wider application in risk assessment. Examples were provided of conflicting plant responses that do not only depend on the given pollutant and pathogen or herbivorous insect, but also depend on the plant species or even genotype and the specific growth conditions. Interactions between pathogens and abiotic factors in plant response demonstrate the need for increased awareness of these complexities, as they affect competition, species composition, and ecosystem functioning.

Subtle biotic interactions within a forest may have more important consequences than the direct effects of pollution. Current knowledge on this issue leads to the assumption that chronic pollutant impacts may weaken plant defense to other stress factors rather than to strengthen it. Such a pathway of pollutant action is believed to be one cause of the successional changes in the $O_3$-impacted forests of California. Long-term changes in the gene pool of tree populations may result in altered species composition with consequences for food webs and fire ecology. Prognoses become uncertain when inter- and intra tree species competition is more important than stress by pollutants and parasites. Pathogen infection may be more sensitive to climate than pollutants. Apparently, ecological scenarios vary in regard to plant and parasite responses, and difficulties in a full understanding will prevail as long as mechanistic explanation, including genetic control, is inadequate.

Guidance for new investigations can be provided by theories about resource allocation such as the Growth Differentiation Balance Theory (GDB). GDB addresses the trade-off between growth- and defense-related metabolism, with the latter being favored at low resource availability, and the former favored at high resource availability. Metabolic control and metabolic activity (i.e., molecular and "organismic" biology) should be studied in parallel; the latter representing the biochemical and ecophysiological process levels of the whole plant. The trade-off addressed by GDB needs to consider also the resource exchange with other pools, such as the mycorrhizosphere. In addition, conventional scaling of processes needs to account for ontogenetic stages and growth conditions (controlled vs. field), with actual pollutant uptake and effective dose having paramount importance. Research outcome needs to validate models as an analytical and integrative tool capable in quantifying allocation trade-offs across scenarios. The ultimate goal is the mechanistic understanding of interactions, implying spatiotemporal integration of resource allocation across the plant-plant, plant-parasite, and plant-mycorrhizosphere relationships, while examining findings for their extent of generality across site conditions, tree life forms, and ontogenetic stages.

**Detection, Monitoring, and Evaluation**

Monitoring is essential to document current status and changes in forest health. Two topics that came to light during the meeting merit further investigations: (1) changes in plant phenology due to climate change and (2) plant competitive response to pollution. The length of the growing season is an important factor affecting forest productivity. Compared to historic conditions, the growing season now begins up to 15 days earlier in central-northern Europe. At the community level, species-rich plant communities were shown to experience more disturbance impacts under pollution compared to species-poor communities. This conclusion contradicts the widely accepted theory of greater resilience of species-rich communities and contributes to the discussion on the relationships between diversity and stability.

$O_3$ was the pollutant of major concern for forest scientists and this reflects the activity and the large amount of work carried out at the international level.

Attempts to study $O_3$ effects on forests were presented including statistical (i.e., correlative) approaches and perspectives on risk assessment. Under low $O_3$ levels, peak concentrations of $O_3$ seemed to have more significant effects on tree defoliation and increment rather than mean or integrated AOT40 values. However, $O_3$ concentrations and exposure-based indices explained a limited portion of the total variance in crown transparency and basal area increment. The flux-based approach in risk assessment was advocated. Many questions remain unanswered including: (1) Is nonstomatal flux an important part of total $O_3$ flux? and (2) Can cumulated flux and cumulated exposure diverge considerably according to the precipitation regime? There is a clear need for coupled long-term monitoring and short-term experiments to translate stomatal $O_3$ flux to biological effects. In this context, long-term studies may provide valuable information. In particular, a better integration among related programs (e.g., ICP Integrated Monitoring, ICP Forests, and national projects in Europe) is needed.

### Action Mechanisms and Indicator Development

Mechanisms of action and indicators of pollution stress continue to be areas of the largest contributions to air pollution research. It is particularly challenging to identify specific or unique indicators for individual stresses. Many biochemical indicators are not pollutant specific, making it difficult to identify causative agents in atmospheres with a mixture of pollutants. Detailed microscopic techniques are being used successfully for diagnosing $O_3$ and other stresses, such as heavy metals, irrespective of the site and species, although certain species-specific responses exist.

Over the last decade, widespread use of stable isotope technology in air pollution research has improved our understanding of the physiological effects of pollutant stress and, in some cases, has been used to reveal pollutant sources. The increased availability of mass spectrometers has driven the development and application of this powerful analytical tool. There is no doubt that stable isotope technology will be an important component of future air pollution research, particularly to assess effects in the field.

Modeling stomatal behavior continues to be a significant challenge as we attempt to move away from exposure statistics and replace them with a more meaningful estimate of actual uptake or dose. Most models and exposure statistics consider only daylight hours when calculating $O_3$ exposure and/or uptake, yet new results show that some species exhibit stomatal opening at night. In some cases, stomata remain open at night after daytime $O_3$ exposure. These responses appear to be transient and reversible, and may lead to sluggish stomatal behavior. In addition, the presence of $O_3$ may influence the response of stomata to light, further complicating estimates of stomatal uptake. Models need to include moisture stress as well as VPD when estimating uptake, and recent efforts to use eddy-flux estimates may lead to improved estimates of $O_3$ uptake. Collectively, these efforts will provide new information that may be important to establish protective air quality standards in North America and critical levels in Europe.

Physiological understanding of plant response to $O_3$ has been studied for decades, yet there is still much that is not understood. The perception of visual leaf injury, night and winter exposure effects, climate and interorganismic interactions, acclimation and adaptation potential of natural vegetation, and the disruption of metabolic pathways are all largely unknown. With the development of molecular tools, such as microarrays, researchers now have the ability to look at changes in gene expression and to examine how stresses, such as $O_3$, may influence plant metabolism at the molecular level. These powerful tools will continue to provide important mechanistic information on plant response to $O_3$ and other atmospheric stresses, and may yield stress-specific indicators of response. However, integration is needed across the molecular and biochemical/(eco-)physiological process levels to develop a mechanistic understanding at the whole-plant level.

It is clear that plant response to multiple stresses and multiple pollutant interactions is one of the most significant challenges facing air pollution researchers. Ozone during the summer has been shown to affect the ability of certain tree species to tolerate freeze-thaw events. Because of the dynamic nature and timing of specific climatic events that lead to this response, it may be difficult to link resulting damage to $O_3$. It was also reported that drought and $O_3$ interact in certain tree species. Again, these responses are species specific, site specific, and climatic-zone specific, presenting a significant challenge for unifying an approach to assessing forest $O_3$ flux- (Europe) or exposure-based (North America) effects.

**Soils and Nutrient Cycling**

Many definitions of N saturation have been offered including some based on the degree of net retention of N deposited to ecosystems. Results from the N application at the Fernow Experimental Forest in West Virginia bring this into focus: while total N retention went up to ~90% with N amendments (suggesting less N saturation), soil solution nitrate levels increased (suggesting more N saturation). Thus, it does not seem advisable to define N saturation on the basis of net ecosystem N retention. Other presentations showed that critical loads for N vary substantially with other conditions, such as drought, levels of other pollutants, and stand age, bringing into question the concept of a single critical load for an entire region. An update on the development of critical loads estimates for N, sulfur, and acidity for U.S. forest ecosystems, and a case study where critical loads for N were dependent on multiple stresses (including drought and insects) suggest that multiple stresses must be included in critical loads analyses.

Although long considered a negative effect, recent studies have shown that soil compaction in coarsetextured soils on drought sites can cause greater plant growth by increasing soil water-holding capacity. Similar to many other factors in soil fertility, it is the degree of compaction and the site conditions that determine its effects, and broad generalisations are not advisable. A second new area of soil science that relates to many air pollution issues is that of "hot spots" and "hot moments", about which several papers have been recently written. These are microsite locations in the soil where nutrient availability is high (hot spots), from which plant roots can take up nutrients that are limiting in the bulk soil. These hot spots are ephemeral, as plant roots deplete them during "hot moments". As these sites move from place to place, plants roots also concentrate in these microsite locations, thus potentially allowing plants to obtain a larger proportion of soil nutrients than may be indicated from bulk soil analyses. This is an area that can change how we view plant root–microbial competition for N and other nutrients, and deserves much more research and consideration.

**Genetic Aspects**

The development of molecular tools that address genomics, transcriptomics, proteomics, and metabolomics is rapidly expanding our understanding of the basic genetic control of physiological traits. These approaches are complemented by the now complete genome sequences for *Populus*. Together with QTL (Quantitative Trait Locus) approaches, researchers have unprecedented opportunities to link variation in complex physiological traits to DNA sequences and to specific chromosomes. This new approach is particularly useful to examine the responses of trees to stresses, such as air pollution and climate change. Studies on DNA methylation demonstrate impacts of environmental conditions on gene regulation and corresponding modifications of the adaptive abilities of juvenile plants of *Picea abies* to climatic stresses.

# 12

# Air Quality Management

### Improving Information

Information is essential for successful Air Quality Management. For an issue specific working group on Air Quality to progress effectively through the Environmental Planning and Management process, a sufficient amount of relevant and reliable information is necessary at each stage. Information must be properly analysed and understood so that technical dimensions of the problems are known and the feasibility of various approaches is appreciated.

Cities have found that information about air quality and related problems is often limited, at least in terms of what is readily available. Useful information and expert knowledge often exists but is not readily accessible. It is typically found in many different forms in many different organisations and groups. These organisations and groups usually do not easily share with one another. In practice, working groups generally find that they do not need to generate much new information. Their real task is to better identify the available information and expertise, bring them together and restructure and reformat the information to make it easily applicable to the needs of the working groups, the Air Quality Management process, and their specific urban situation.

Despite the highly technical nature of some of the information, it is generally possible to create a user-friendly form relevant for Air Quality Management activities.

The first tangible output of the EPM process in a municipality usually is the development of an Environmental Profile (EP). Key stakeholders from different sectors – private, public, and community – are involved through a consultative process in the preparation of the Environmental Profile and in identifying the environmental priority issues facing the municipal areas. The main objectives of the EP are to clarify the environment-development interactions in an urban setting. A City Consultation follows the preparation of the profile where stakeholders from all levels of government and relevant sectors come together to deliberate and agree on environmental priority issues confronting their local authority. The Environmental Profile focuses on the environment from an urban development point of view.

**Environmental profile**

Experience with City Consultations has shown that air quality is a priority issue for many municipalities, cities, and agglomerations in developing countries. The EP provides an important information base because it serves as a common context for all groups working on air quality and other environmental issues. By design though, the Environmental Profile is not meant to contain very detailed information on any one resource (such as air). Working groups have generally found that more detailed information is necessary, because the issues must be further clarified before considering options and formulating strategies. Consensus is usually reached on the need to prepare an Air Quality Profile (AQP).

Since the Environmental Profile (EP) covers all the city's important environmental issues, the Air Quality Profile will contain valuable information on air quality and the factors affecting it (such as activities with negative effects, conflicts of interest over air pollution, overview of the organisations and groups involved in air quality management). The EP provides the first basic source of information for any air quality working group (or other groups grappling with air quality management issues). The first task of any working group preparing an Air Quality Profile should be to thoroughly review the EP in order to extract the wealth of available information and to identify gaps where progress needs to be made.

Developing country cities have found the following useful ways of improving the air quality information base:

— Preparing an Air Quality Profile

— Detailed mapping of the city's air quality situation / problems

— Holding a city-wide consultation on air quality issues (a mini-consultation)

***a) City air quality profile (AQP)***

The Air Quality Profile (AQP) is a document similar to the general Environmental Profile (EP) but is focused exclusively on air quality. It follows the same structure and logic as the EP. The Air Quality Profile's objective is to discuss, in detail, the present air quality situation in the city, to examine how and which activity sectors are causing which kinds of air pollution, to identify those affected by air pollution, to show what the specific problem areas are, to highlight conflicting interests, and to assess the existing policies and institutions active in addressing the issues.

***b) Mapping***

Monitoring data on air pollution (by type), location of specific air quality problems and main polluting sources, concentration of respiratory diseases, distribution of citizens complaints, etc are all relevant information that can be plotted in maps. Putting the information into maps is a helpful analytical device that clearly shows spatial relationships between pollutants and human activities. Detailed maps should be an integral part of an Air Quality Profile.

***c) Consultative meetings on air quality***

As the air quality working groups discuss the topic of air quality in greater detail, they begin to see the need for participation of a larger group of members. Increased participation can be accomplished through specially organised workshops, often called mini consultations. These are organised similar to the more comprehensive City Consultations but with a specific focus on the issue of air quality. Broad stakeholder involvement in these consultations remains very important.

Cities have found mini-consultations useful for:

— Creating and exploring new information or insights, which had not yet been included in the city air quality profile;

— Providing an opportunity for stakeholders representing different interests to give their opinions about the information collected for the Air Quality Profile;

— Giving a much clearer idea of what are perceived as the most urgent air quality problems, for whom, and why, through broad-based participation;

— Supplying a variety of useful inputs, which can be used to update and revise the Air Quality Profile;

— Discussing strategy ideas and proposals in order to obtain initial reactions and to elicit more ideas and options;
— Forming further working groups on specific sub-issues related to air quality due to the wide variety of stakeholders present.

**Involving the Stakeholders**

Cities all over the world have found that involvement of a broad range of city stakeholders is essential for successful air quality management. Wide-ranging participation by all interested groups and organisations is vital for any type of urban environmental or development management process. Stakeholders must be involved at all stages of the process including: information collection and analysis, prioritisation of issues, review and assessment of strategies, formulation of strategies, action planning, and most importantly implementation of investment projects and institutionalising the entire process. Cities have been successful in addressing air quality problems by bringing in a wide range of relevant stakeholders. These stakeholders include:

— Those who possess expertise in air quality management and/or have important information about different aspects of the issue;
— Those whose interests are directly affected by urban air quality management issues;
— Those who control or influence important instruments or mechanisms of management;
— Those whose actions directly impact on the air quality situation.

One of the early tasks in a SCP project is to systematically identify the stakeholders who are relevant for the various priority issues. One may identify stakeholders by starting with the knowledge possessed by the core participants in the project. Additional stakeholders can be progressively identified through the preparation of the Environmental Profile and through the preparations for and holding of the City Consultation. There is a major effort involved in the early stages to find and mobilise stakeholder participation. As the work proceeds, the range of participating stakeholders is continually refined and supplemented.

One important reason to involve a wide variety of stakeholders is to facilitate the assembly of relevant information. For example, universities, various governmental research departments and institutes, industries, and other organisations often have and/or collect air quality data on the city. However, this information often is incompatible and recorded or kept in different formats because the information was collected for different purposes. Other information relevant to air quality (e.g. motor vehicles and traffic information) is available from other organisations. Quite commonly, the separate organisations have little idea of what data is gathered or accumulated by the other. There is generally very little communication and almost no exchange or sharing. Some data will often overlap while other, important data, is missing. Thus, cities often find that information needed for air quality management activities is not readily available.

Working together on an agreed set of issues and problems can foster a willingness to share data and information, as many working groups in SCP demonstration cities have discovered. A related function that the working group can fulfil if its membership is sufficiently inclusive of key stakeholders, is to help integrate the different and perhaps incompatible sets of data and information. This process, while difficult and time-consuming, can help create a reasonable information base withcut the considerable expense attendant to new primary data gathering. Also, a broad-based working group membership can help identify the gaps in the various data sets so that measures can be taken to find missing information.

## Clarifying Issues

Conflicts of interest between stakeholders over air quality issues do happen. In order to formulate guided strategies and action plans, it is imperative that air quality issues and their relation to development activities be carefully analysed and clarified. These clarifications can be achieved by applying all available technical means and by broad stakeholder involvement.

### *Involving stakeholders in clarification*

Air quality issues can be clarified with as much detail as possible through many means including the preparation of an Air Quality Profile and mapping exercises for air quality purposes in the city. Depending on the level of information obtained from these various means, working groups with representatives from a broad range of stakeholders can be constituted. The task of the working group is to further clarify the issues that require more definition and analysis. When clarifying issues, working groups should first focus on:

— Identifying and collecting the existing data on the issue of priority;

— Converting and presenting this data in a format that is useful for the working group;

— Identifying missing information on the issue of priority;

— Identifying the appropriate sources of missing information, i.e. who is likely to possess this information;

— Developing strategies for obtaining the information as well as linking up with these potential information sources.

Information can be available from various government departments, research institutes, universities, private sector organisations, and popular sectors working in air quality or general environment and development related fields. A deliberate attempt should be made to include these informative stakeholders in the working group.

The work of this group may entail field visits to and surveys of, for example, mobile stations to witness and experience the level of pollution and to register complaints by various affected persons. One of the main responsibilities of the working groups is to collect the scientific and technical data available about the city's air quality situation and problems, and translate this data into usable information for the use of the working group and for the city stakeholders. Many cities have monitoring equipment and systems for specific air pollutants. Many cities know how much traffic of which types is going through which areas. Many cities have data on the type and amounts of emissions by industries, etc. Often, this data is collected and stored in institutes and technical reports that do not reach the desks of decision-makers, let alone other stakeholders. Even if the information does reach the stakeholder, it may not be in a readily usable format. Working groups should make the effort to determine what exactly is available in the city, which organisations are involved, and which organisations should be a part of the working group. It is important to be able to identify the right sources for information needed to further clarify an issue. It is equally important that this exercise produces all relevant information necessary for the working group to be able to follow the EPM process and come up with strategies and action plans.

The involvement of technical experts helps clarify and prioritise issues. Their technical expertise and experience bring credibility to the process. They not only enrich the information available but also are instrumental in facilitating the process of consensus building. However, the role of experts is subordinate to that of the stakeholders. Even with the best experts around, not much can be done without the support of the stakeholders.

*Emissions assessment for clarifying issues*

Emissions assessment techniques are an indispensable part of air quality management. With technical experts available, emissions assessments should be undertaken because they help identify the main sources of pollution in the city and assess the interconnectivity between the various sources of pollution. Emissions assessment also helps to identify the underlying reasons for the polluting activity sectors. Emissions assessment illustrates, for example, that factories equipped with (often simple) pollution control measures emit less than uncontrolled factories. pollution control equipment for these brick industries and improved the air quality by reducing Suspended Particulate Matter (SPM).

There are many ways to further clarify air quality issues. The following methods are discussed below:

a) Monitoring the pollution
b) Calculating/ estimating air pollution
c) Use of mapping
d) Emissions inventories

*a) Measuring the pollution by monitoring*

Monitoring air quality provides the necessary baseline information to identify which pollutants are of major concern and are principal sources of pollution.

Continuous monitoring is necessary to clarify air quality issues in greater depth and certainty. The reliability of the monitoring is critical. The methodology used for monitoring must be sound. Actual monitoring must be well documented. Both low-tech and high-tech methods of monitoring are readily available and each one has its advantages and disadvantages. Whichever technology is chosen must be appropriate in terms of the technical capability of the users and the availability of the resources to operate and maintain the technology.

In order to develop a focused monitoring programme meant to further clarify issues (and later to assess improvements due to strategies adopted and implemented), there should be an initial assessment of the pollutants when you select and prioritise air quality issues within the city. Some primary data (such as number of cars within the city, $SO_2$ or SPM levels, emission inventories) that may have been prepared is often available but scattered. This initial assessment will prevent problems (i.e. monitoring pollutants that are not problematic within the city itself). The experience of many cities, long-term commitment, long-term funding, and training are key factors of a successful monitoring programme. There are basically four ways to measure urban air quality: 1) passive samplers 2) active samplers, 3) continuous analysers, and 4) remote sensors. Many cities measure the quality of the city's air through monitoring stations. These stations measure the concentration of pollutants in the air, usually including $SO_2$, $NO_x$, $PM_{10}$ or SPM, VOCs, etc. The measurements of these air pollutants can be done continuously or on a regular (e.g. weekly) basis. The different sampling stations can either work independently or can have direct connections to a central station. This information can be displayed in a time series, showing the change in concentration of the pollutant over time.

In addition to fixed monitoring stations, some cities have mobile stations. These are fully equipped vehicles that can either assess the air quality at any place, or take samples to be analysed later. These mobile stations are often used to investigate complaints and measure the compliance of industries with existing regulations. Information obtained from these stations is particularly useful as a close correlation between source and effects of the pollutants can be established. The combination of information available from the monitoring stations and those from the mobile units can be better presented through preparation of air quality contour maps.

*b) Calculating / estimating the air pollution*

Measuring air pollution can be very expensive. Not all cities have extensive monitoring networks and/or mobile monitoring equipment. Models and calculations are increasingly being used in the case that this kind of equipment is scarce or the capacities to appropriately analyse samples is weak. Computer models are a relatively fast and inexpensive way of providing air quality information. Because models can be used to evaluate air quality management options, they are also suitable for planning and strategy development.

For example, the average background concentration within a municipal area caused by vehicle exhaust can be calculated if parameters, such as the number of cars, the average mileage driven within a day and the volume of exhaust produced are known. The CAR International Model can assess pollution levels of certain streets.

The cost-effectiveness of policy measures can be assessed in advance and/or monitored during implementation. If, for example, the emissions of different types of vehicles are known, the achieved emission reduction (when a certain reduction of vehicles has been realised) can be better assessed. Emission calculations make clear what the most serious polluting activities are. This allows for intervention and setting up cost-effective strategies and action plans. It is therefore strongly recommended that emission assessments, like air quality monitoring, be carried out on a regular basis. Emission assessment activities should be institutionalised and should include the exchange of relevant results with the decisionmaking actors/institutions.

Predicting air quality from multiple sources by using modeling is very complicated. Largescale models that can handle multiple sources and different pollutants are used only by very specialised organisations. Air Quality Models are best used for isolated sources or situations. For example, Air Quality Models may be used on air pollution caused by traffic (one specific source) in a certain street (one specific place). Air Quality Models may also be used for predicting the air pollution caused by an industry stack, based on wind conditions, how the plume of the stack goes, where the pollutants will be found, and in what concentration.

Advantages of calculating concentrations with dispersion models:

— It reduced the need for complex (and expensive) ambient air monitoring. However, monitoring remains important to calibrate the dispersion models.

— Models can be used to assess the cost-effectiveness of policy measures in advance. By varying the input data one can assess beforehand the expected air quality improvement of response options. For example, the "CAR International" model models various options of interventions (such as re-routing of certain types of vehicles) and can calculate the concentration of pollutants respectively.

– To check future air pollution caused by new factories as part of an environmental impact assessment, it is recommended that dispersion modeling be carried out early in the project preparation, before the plant location and detailed design have been finalised.

*c) Using mapping for clarification*

Mapping is also a very useful tool to display information and clarify issues. Thematic maps will be useful in analysing the extent of the pollution problem from each activity sector.

*d) Inventory of emissions*

Preparing an emission inventory, is a compilation of all air polluting activities in an area. It has two main components: 1) the pollutants (e.g. $NO_x$, $SO_2$) and 2) the sources (e.g. industry, traffic, and domestic). It specifies the location of each source (point source or mobile source) and the time variations in the emissions. Emission inventories can be of great assistance in clarifying air quality issues because they, among other things, assist in the evaluation of emission trends, which in turn assist in formulating air quality management policies. The construction of a complete, high quality emissions inventory is, however, time consuming and complex.

***Clarifying city/area - wide urban air quality problems vs. 'Hot spot' urban air quality problems***

Further clarification of the city's air quality issues may require an analysis of the problems at different levels within the city. On the one hand, a clear understanding of issues resulting from individual industries or traffic sections, and how they impact their vicinity, may be required. On the other hand, clarification may be needed on the overall impact in the city by the various activities.

*a) City / area-wide urban air quality problems Transport*

Transport is a main contributor to high concentration of pollutants in the atmosphere. In many cities, traffic, sometimes in combination with industrial air pollution, is responsible for smog. Cities are increasingly experiencing these problems. On hot summer days, the ozone level may exceed the norm and warnings are given. Emergency measures may be taken such as car bans on particular days. People with cancer and asthma, the elderly, and children may have to stay indoors during such times.

*b) Clarifying site specific urban air quality problems (Hot Spots) Transport*

Air pollution in specific streets depends on the volume of traffic and on street construction factors, i.e., distance to houses, height of houses, and trees among others. Pollution can be calculated and, if possible, verified with random measurements and included in a city map. The CAR model computes the concentrations of $NO_2$, benzene, and CO at predetermined points in the street. The input data are street geometry, information about the traffic (mean daily traffic and composition), and the background concentrations for specified pollutants.

IMMIS LUFT, which runs on the same principles as the CAR model. However, IMMIS LUFT also includes the air pollutants HC , soot, and $CO_2$. It is useful in calculating the annual average concentrations of benzene, carbon, and the 98 percentile value of nitrogen. The model is used in many German cities. The advantage of this model is that it can be used in combination with GIS (Geographic Information Systems) for mapping. It can also be used with the transport planning system

VISUMin order to estimate the effectiveness of transportation planning on air pollution.

***Industry***

Tools for calculating concentrations near polluting activities are based on dispersion models. A dispersion model is a software programme that calculates the concentration of a specific pollutant by using the emissions from the various activity sectors. Dispersion models may give the same results as ambient air monitoring in the cities. Both may give the concentration of a specific pollutant in $g/m^3$ air. The dispersion models are especially useful to analyse air pollution near industries.

***Indoor pollution***

Compared to outdoor sources (such as factories, power plants, heating and vehicles), indoor air pollution is not a large contributor to overall emissions. Nevertheless, a person's health can be seriously affected by exposure to intense indoor air pollution. Inappropriate burning of fuels, bad ventilation conditions, as well as long exposure times are responsible for illnesses, such as respiratory diseases. In traditional societies, particularly women, who spend most of their time in house, are affected by in-door pollution.

Much research has been done on the indoor air pollution emanating from burning fuels. For example, there is an indoor air pollution database for China. It is based on data prepared for China that takes into account pollutants, fuel type. The adverse effects on human health can be pinpointed by comparing the results of the assessment with the air quality guidelines, and improvement of in-door air quality can again be assessed after having implemented response options.

**Prioritising and Selecting Air Quality Issues**

Almost always, the experience in the SCP demonstration cities has been that air quality issues were not well defined at the onset. People may think that they 'know the issues' but in fact only have a broad outline of the issues. In general, there is a considerable degree of uncertainty as to what exactly is meant by a particular issue: what is really the problem, for whom, why, and in what way? A broad topic like 'air quality' may be comprised of a wide range of sub-issues. Air pollution may mean many different things. More often than not, it is necessary to spend considerable effort to identify and prioritise the particular sub-issues that will become the focus for the air quality working group.

After having met a couple of times, some working groups found that the main issues were straightforward (like pollution from a certain industrial area, from traffic in the city centre, or from small-scale industries in specific locations). In these cases, the working groups could select well-defined sub-issues and take them as priorities for further work. The working groups may decide to set up a series of sub-working groups, each focused on one of the priority sub-issues.

Cities have used a variety of methods to identify and prioritise issues. The criteria listed in the EPM Sourcebookhave proven useful in many cities. These are systematic criteria helping cities to - more confidently and more transparently - make decisions about which priority issues or sub-issues to work on first. These criteria include:

a) The magnitude of the health impacts associated with the problem
b) The size of urban productivity loss caused by the problem

c) The relative impact of the problem on the urban poor

d) Whether or not the problem leads to an irreversible effect

e) Whether special circumstances offer special opportunities

f) The degree of social/political consensus on the nature or severity of the problem

g) Whether the problem is city-wide or specific to a "Hot Spot"

The relevance of some of these criteria for prioritisation of air quality issues and sub-issues is discussed below:

***Magnitude of health impacts associated with the problem***

The health effects of air pollution have always been one of the major reasons for focusing on air quality management. To better understand how different air pollution problems affect health, a number of approaches can be taken. WHO guidelines on air quality are a very useful reference – especially for countries with no national guidelines. WHO guidelines are widely accepted and scientifically sound indicators of the levels of pollution exposure. The potentially harmful effects of exceeding these standards are also given in these guidelines.

The health effects of air pollution are highly variable. Different pollutants have different types of health impacts. Health impacts vary by concentration and by type of exposure. Air pollution encompasses a range of different types of contaminants. It is extremely important to understand these different effects. While detailed knowledge is not necessary at this stage of work, a broad idea of the health effects of air pollution is extremely important.

***Amount of urban productivity loss created by the problem***

In general, translating environmental pollution into economic degradation is more difficult than assessing health impacts. Although the science of environmental economics has developed ways of assessing economic damage resulting from environmental degradation and depletion, this methodology remains somewhat difficult for non-experts. This is particularly useful with respect to dust (in this case meaning all suspended particulate matter of whatever size) because illnesses resulting in lost days of work are one of the greatest impacts of high levels of total suspended particulates (TSP). In the long run, for example, days lost to education through sickness in children will have a negative effect on economic growth. Other economic losses could include reduction in a city's tourism industry with visible air pollution acting as a powerful deterrent to tourists, and damage to historic and cultural landmarks by air pollution causing additional costs of repairs and maintenance.

Persistence of high levels of air pollution can also have an important disincentive effect on economic development. The general impression of a city is one important factor influencing investment. In general, cleaner cities attract more external investment. Some industries will not consider locations in areas of high air pollution because of production requirements. For example, high-technology electronics manufacturing is sensitive to atmospheric conditions. The protection of production processes against heavy air pollution can be prohibitively expensive.

***Relative impact of the problem on the urban poor***

Urban air pollution has a relatively higher impact on the urban poor than on the general population for several reasons:

- The health of the poor is often below average, reducing their resistance to disease and increasing the chances that they will suffer health effects from air pollution;
- The housing of the poor is usually low in quality, badly ventilated, heated by basic systems using fuels and techniques that produce high levels of indoor pollution (in some urban poor areas, indoor air pollution is the most serious health threat);
- The urban poor often live in less attractive areas near air pollution sources in heavily exposed downwind areas. This typically exposes them to high concentrations of air pollution that are much more severe than the average city levels. The Environmental Profile usually includes information about where the poor live and where the lower-quality settlements are situated in the city. This information combined with the more detailed information in the Air Quality Profile should provide a reasonable picture of how and where the poor are affected by urban air pollution.

***Whether or not the outcome leads to an irreversible***

Effect When assessing air pollution effects it is important to distinguish between reversible and irreversible outcomes. The air in most cities is replaced within a matter of days; if motor vehicles and industries stopped producing polluting emissions, the quality of the ambient air would improve dramatically and swiftly. In contrast, the contamination of soils or ground water tables can often be reversed only very slowly, if at all. There may be many irreversible impacts even if the levels of air pollution drop dramatically. The corrosion of buildings is one such example. If a building is exposed to serious air pollution for long enough, the stonework or structure may be damaged to a point beyond repair. Acid rain, a precursor of acid forming pollutants present in the atmosphere, also has grave effects on historical and cultural heritage. Acid rain caused by $SO_2$ and $NO_x$ pollution in the city dissolves soft stone, like limestone. This can cause damage to old buildings, statues, and other exposed materials. Besides being a continuous threat to forests, lakes and soils, in many old European cities acid rainis a real threat to old buildings, monuments, and statues.

*Whether special circumstances offer special opportunities*

Special circumstances in a city may provide very rare opportunities for addressing air quality situations even if air quality had not been a priority before. In certain cases, air pollution may be addressed because of its strong relationship to other problems being addressed by city management. For example, if the city managers were attempting to reduce traffic congestion within the city centre, the application of public transport and traffic strategies goes hand-in-hand with air quality management in the city centre.

Another example is spatial planning. Spatial planning measures undertaken by city planners could improve the air quality situation of the city. Planning undertaken to mark zones for the operation and expansion of industrial and business activities, away from residential places, helps improve the air quality of the city and reduces the exposure of the city's population to hazardous pollutants.

Special events of a regional or global nature can act as a catalyst to promote local action on air pollution. For example, the global concern about the effects of climate change culminated in the United Nations' Framework Convention on Climate Change (UNFCCC) and in the United Nations' Kyoto Protocol. In order to meet obligations under the convention, governments first sign and ratify the convention and then act locally. Many of these actions directly address air pollution issues.

Air pollution episodes, accidents and disasters also create change. These accidents turn the spotlight on air pollution problems and thus offer a valuable opportunity for including air pollution as a priority. They ignite radical air quality management strategies. Some examples include:

-- Explosions of factories, or gas leakages such as the one that occurred in a Union Carbide factory in Bhopal, India, shocked the whole world. In this accident, the leakage of the highly toxic gas methyl isocyanate (MIC) killed thousands in 1984. Though it reached an out-of-court settlement with the Indian government, Union Carbide refused to accept responsibility for the disaster, blaming it on terrorism and industrial sabotage. However, strong public pressure forced Union Carbide as well as other large chemical companies to implement stricter safety and environmental standards.

— The release of highly toxic dioxin from a small chemical plant in Seveso, Italy, in 1976, led to harsh industrial regulations issued by the European Union.

— The oil crisis in 1974 caused a new perception regarding the use of fossil fuels. Many countries in Europe introduced a so-called 'car free Sunday' that prohibited normal car users from using their cars on Sundays. This policy measure, while introduced to curb fuel consumption, had a positive effect on air quality and focused attention to this issue. Today, Access is closely involved in actively promoting the European Mobility Week - including the Car-Free Day on 22 September each year.

— Excessive congestions and high concentration of air pollutants in the city centre of London caused the introduction of a radical 'congestion charge'. This resulted in a drastic change in modal share towards public transport. Other cities introduced different measures to restrict access of private cars to the centres.

***Degree of social/political consensus on the nature or severity of the problem***

Many times, identified air pollution problems need to be accompanied by sufficient scientific data for further clarification. However, in certain cases, the effects and presence of air pollution are very obvious and there is common concern and consensus for action. In such situations, the priority is to act immediately and to not delay urgently required counter strategies, especially if the delay is due to lack of scientific data. At this point, the degree to which the pollutant(s) exceed the laid-down national standards or WHO standards may not be relevant. Complaints arising due to health problems and other complications should be sufficient to trigger response from city managers and decision-makers. The citizens' health problems and complaints should be enough to make the city act. This is particularly true as air quality problems can become manifest in different ways. Air quality problems may be very local and particular to a spot in the city and impossible to measure. In such a situation, the average emission levels in the city may be below standards, but because of local conditions (e.g. strong winds always transporting air pollution towards one direction or the existence of inversion layers) citizens might experience severe problems. The stakeholders, i.e. those affected, should be allowed to be part of the problem definition and solution search.

Political and social consensus on a problem is very important. It is important that all stakeholders are involved in the process of preparation and decision-making to make sure the city and those responsible address al. air quality problems. Air pollution also has social costs. Citizens want to live in a city with clean air. Studies based on willingness to pay for clean air suggest that the social costs of air pollution in cities even exceed the cost for direct damage to health and property.

***Whether the problem is city-wide or specific to a "Hot Spot"***

Distinguish between city-wide or area-specific (hot spots) air pollution. Examples of citywide air quality problems include smog caused by traffic in large areas of the city and citywide background air pollution concentrations caused by large industrial areas in or adjacent to the city. Climatological conditions like inversion layers, wind directions, and geographical conditions in a city can affect the concentration of pollution and the magnitude of its impact. For example, hills surrounding a city exacerbate the air quality situation. On the other hand, other natural conditions specific to a city can help reduce air quality problems. The air above a city is able to refresh itself within one-half to three days depending upon the wind, altitude, and absence of geographical barriers (such as mountains and hills).

Hot Spots are specific problems in specific locations. They may include an industry polluting its direct vicinity, pollution of the city centre, and small-scale industrial activities taking place in urban areas. These medium-scale problems are often not affected that much by weather and geographical setting. A main issue in this category is the risk of air pollution caused by industrial accidents. These issues can be prioritised by working groups formed for specific problems in certain areas/Hot Spots within the city and for city-wide air quality problems.

## Improving Strategies, Action Planning, and Decision-Making

While the previous chapter discussed how to acquire information that is necessary to clarify the problems and identify the priority issues, this chapter discusses how cities prepare, assess, build consensus for, and reconcile air quality management strategies and action plans. It includes suggestions on how cities can incorporate/integrate these strategies into existing urban management strategies and policies for better action planning. Volume 4 of the SCP Source Book Series on Formulating Issue-Specific Strategies and Action Planscan guide this step of the process.

### Formulating and Clarifying Air Quality Management Strategies

To be able to start developing and comparing strategies effectively, distinctly defined issues have to be addressed. When developing and comparing possible strategies, it is recommended that working groups take the following points into consideration:

a) Pros and cons of each option for various stakeholders;

b) Potential air quality improvement of each response option;

c) Social and economic costs and benefits of each option;

d) Associated policy reforms and institutional strengthening that have been agreed upon to support implementation of the strategy;

e) Long-term environmental targets and interim goals to guide phased interventions;

f) Indicators that can be used to track the progress of actions and their impacts;

g) Cost-effectiveness analysis and who bears the cost of implementing a strategy;

h) How to implement the strategy;

i) The success of multiple instrument strategies;

j) Opinion and reality check – what is already happening in terms of air quality improvement and how do people perceive the problem?

***a) Considering the pros and cons of each option for various stakeholders***

City experiences have shown that the acceptance and implementation success of strategies depends on the reconciliation of gains and losses in a fair and equitable way. A situation where all stakeholders win must be created. Any strategy developed or improved should be through an integrated approach to the problem being addressed. It is important to realise that air quality issues involve different decision-making sectors. These crosscutting issues deal with legislation, monitoring, economic policies, traffic policies, industrial development policies, energy development policies, land use planning, and others.

A strategy should be developed in cooperation with the stakeholders, both within the government (national, regional, and local) and those outside the government (the private sector and civil society). Involving all stakeholders in the strategy development process ensures that due consideration is given to the various interests of the stakeholders, including the pros and cons of a particular strategy from their own perception. This will promote reconciliation and consensus on issues and strategies and may lead to a sense of ownership over the chosen strategies among the stakeholders.

***b) Assessing the potential air quality improvement of each response option***

In selecting the response options or strategies for addressing air quality issues, it is very important to assess the potential effects of the strategies. Tools that can be used are emission assessment tools and dispersion models assessing air quality near polluting activity sectors. Knowing how much improvement can be expected in implementing a strategy against other strategies that address a particular air quality issue promotes consensus on the strategy.

***c) Considering the social and economic costs and benefits of each option***

When developing strategies, it is essential to predict the effects of implementing these strategies. One must determine the effectiveness and efficiency of each strategy to address the air quality issue. Due consideration should also be given to possible side effects of the strategies, particularly economic and social side effects. When trying to reduce car use in a city centre, banning cars might be effective, but may have economic and social side effects. Some direct small scale effects of car bans may include reduction of customers in the business centres and scattering of demand flow from between days with car bans and those without.

***d) Clarifying associated policy reforms and institutional strengthening that have been agreed upon to support the implementation of the strategy***

It is important to consider a strategy that will achieve long lasting effects, rather than shortterm solutions. Developing such strategies in a participatory manner, for example through cross sectoral working groups, will help to overcome institutional barriers. Institutionalising such processes at the local level and the demonstration of its successful decision making power can lead to policy reforms at the national levels.

***e) Agreeing on long-term environmental targets and interim goals to guide phased interventions***

No strategy will change the situation overnight. Often a phased introduction is advisable. This could be a step-by-step implementation of the complete strategy or the initial introduction of the complete strategy in a certain area and replicating it in other areas. The strategy should be well programmed and should have a clear time frame for the different phases of implementation.

It is important to realise that people have expectations of strategies. It is highly advisable to develop strategies that can show results in the short-term while addressing long-term issues. If a strategy is addressing an air quality issue only showing long-term results, stakeholders might decide to discontinue implementation because of lack of immediate tangible results.

Experience in the Sustainable Cities Programme has proven that it is important to have an issue-specific working group to develop a comprehensive strategy with short-term and longterm components. Immediate improvements from the short-term implementation provide the support for the long-term implementation of the strategy.

***f) Discussing the indicators and monitoring that can be used to track the progress of actions and impacts***

When designing a strategy, its implementation has to be monitored. It is important to clearly present the different phases of strategy implementation through maps that show the different stages of implementation, documentation of activities, progress reports, etc. In this way, the progress of activities can be tracked and the strategy can be effectively evaluated. Monitoring and evaluation are vital for decision-making and a possible replication of the strategy.

A number of factors must be built into the design of the strategy including verifiable indicators and means of verification. For example, the decrease in the monitored level of total suspended particulates (TSP) in the air is a very good indicator of the impact of a strategy to lower the sulfur content of diesel and fuel oil.

***g) Considering cost-effectiveness and who is responsible for the costs of implementing a strategy***

The cost-effectiveness analysis (CEA) is important for analysing the costs, benefits, and effectiveness of policy and technical measures combating air pollution. The primary objective of this tool is to assist working groups in identifying strategies for attaining a given pollution goal at the smallest cost. A stepwise explanation is given on how to execute a CEA. A city example is also included. Another factor that should be considered is: who will bear the costs of the strategy? Ideally, those that are responsible for causing the problem/pollution should pay, at least for the most part. This is called the "Polluter Pays Principle.

***h) Consider how to implement the strategy***

The success or benefit of a strategy depends on its design and even more on its actual implementation. During the formative stages, working groups should take into consideration factors enhancing the chances for successful implementation of the strategy. These factors encompass public acceptance (which often depends on the early involvement of the stakeholders), adequate capacity or need for capacity building for implementation, resources, and others. The factors must be carefully incorporated into the design of the strategies.

***i) Multiple instrument strategies are more successful***

For a strategy to be comprehensive, it is important to use, or at least to consider all of the possible management instruments available. These include, for example, economic instruments, legal instruments, or communication instruments. Experience has shown that strategies using multiple instruments are most successful. This tool distinguishes between policy/legal instruments, economic instruments, and communication instruments; elaborates on the applicability of instruments; and gives examples of urban air quality management. Practice shows that strategies applying a combination of different types of instruments are most successful in addressing a particular air quality issue. Take the example of reducing motor traffic in a city centre (traffic management). A potentially successful strategy subsidises buses and improves the infrastructure for walking and bicycles. Further, a public awareness campaign would promote car-pooling, as well as the use of public transport or bicycles. A combination of these sets of economic and communication instruments will be more successful for long-term management of the issue.

***j) Opinion and reality check – what is already happening in terms of air quality improvement and how do people perceive the problem?***

When identifying and incorporating existing actions and initiatives within the city, it is important to make sure that the underlying issues are well understood among the population of the city: What do the locals experience as their biggest air quality problems? What are the opinions of resident representatives and resident organisations? What actions are already being undertaken to address air pollution? What strategies/ technologies are locally available and viable?

An example of adaptation is the World Bank approach in Latin American cities, particularly in Mexico. Here, the multilateral agency is sponsoring the development of modern tropicalised tools to better understand and adapt the technologies to local circumstances. This is accompanied with measures like increasing awareness on the scale of air pollution, sponsoring modern Cost Benefit Analysis tools to support decision making and promoting a process of adoption of incentives combined with strengthened regulatory mechanisms.

## Strategies for Different Activity Sectors

According to these activity sectors, air quality management strategies can be subdivided into four categories: transport, industries, natural sources of air pollution, and in-house activities. There is an appreciable use of many strategies and many examples can be drawn from real life experiences. Because of these experiences, some of the advantages and disadvantages of these strategies can now be listed.

### *Strategies to reduce air pollution from transport*

A number of options are now available and are being used in cities to counter the air pollution resulting from transportation. These include strategies for reducing vehicle pollution, for managing travel demand, and for improving the transportation supply. Other transport problems such as congestion, traffic accidents, and social inequities are also taken into account. Cities can use this tool as a quick reference for selecting strategies. The tool also includes advantages and disadvantages of these strategies as well as examples. Strategies for managing pollutants from transportation activities can be categorised as follows;

*a) Reducing vehicular pollution*

Although technical measures alone are not sufficient to ensure the desired reduction of urban air pollution, they are an indispensable component for any cost effective strategy for limiting vehicle emissions. Fuel and the vehicle types have a great impact on air quality situations. This is especially true in many developing countries where the growth rate in private vehicle ownership is higher than in developed countries. Cities in developing countries also have large numbers of older vehicles that were cheaply imported and/or passed down the economic chain.

Overall strategies to reduce vehicular pollution may include:

— *Vehicle inspection and maintenance (I&M):* Inspection and maintenance programmes can successfully reduce emissions from old vehicles and ensure that new vehicles remain in good condition. Pollutants such as carbon monoxide (CO) and hydrocarbons (CnHn) of individual vehicles can be reduced by up to 25% through strict I&M programmes. These programmes accelerate the disposal of old and inefficient cars. However, these programmes may at times face financial, political and enforcement difficulties.

— *Improving fuel quality:* Improving fuel quality in most developing countries involves the reduction of the contents of substances like sulfur and lead. Using more volatile diesel addresses the problem of black smoke from heavy diesel-powered vehicles, such as buses and trucks. Improving fuel quality should also involve the introduction of alternative fuels as, for example, compressed natural gas (CNG) which does not contain lead or sulfur, and TSP (total suspended particulate), and is lower in NOx, SO and CO than conventional fuel. Other options are bio fuels from crops, used cooking oils or bio gas from sewerage and waste. For example, Delhi, India, has converted its entire public transport fleet to CNG and operates the world-wide largest bus fleet (more than 1200 buses) on this alternative fuel. In a further step, all motor rickshaws and private taxis were converted to CNG. The result is that the traffic related air pollution could be reduced considerably. The operating costs are lower than with conventional fuels due to lower fuel prices.

— *Introducing new vehicle technologies:*New vehicle technologies are available and research is continuing. These include promoting the use of three-way catalysts that can reduce emissions up to 90% per vehicle or particulate filters for diesel vehicles. Other types of vehicles run on electricity and the first prototypes of fuel cell powered cars are under testing. But those only fill a small niche market. Hybrid technology will take an intermediary role before switching to new technologies, such as the hydrogen economy. Installing pollution control equipment like particle traps in vehicles or switching from two-stroke to four-stroke engines that allow the use of catalytic converters, help to reduce pollution.

In developing countries, setting strict standards for newly imported vehicles may help to reduce the already existing problem of vehicle pollution in the city.

*b) Managing travel demand and improving transportation supply*

Cities cannot continue to expand their urban transportation systems forever. Although some expansion is necessary, the economic, social, and environmental costs of doing so are prohibitive. Instead, cities must re-examine urban transportation demand and devise new strategies that provide maximum access at minimum total cost. The challenge is to expand and improve supply so that automobile transport becomes one part of the transportation system rather than the focus. However, implementation options for discouraging overreliance on privately owned cars will not work unless people are given an efficient and economic transportation alternative - whether bus, light rail, subway, ferries, walking or cycling. Therefore, improving urban transportation systems will require a combination of policies that reinforce each other and help to avoid adverse side effects. Likewise, the policy to encourage a shift to cleaner alternative fuels should be accompanied by an appropriate pricing policy.

***Formulating strategies to reduce air pollution from industrial sources***

The UNEP report on 'Air Quality Management and Assessment Capabilities in 20 Major Cities'mentions the main strategies to address industrial urban air pollution. The focus is on cities in developing countries. In order to have enough of a basis to formulate strategies that will reduce industrial pollution within the city, it is imperative to have enough information. One would need, for example, information on polluting industries, what type of pollutants are emitted from these industries, what are the trends of the pollutants, and what is currently being done to reduce this trend. It is also important to know if there are any industrial hotspots located within the city. Most of this information should be contained in the city air quality profile developed in the first stage of the process. Another way of obtaining information is through emissions inventories. Emissions inventories are useful quantifications of pollutant emissions over time. They enable better targeting of emissions controls on major sources and source categories. The main strategies for addressing industrial pollution have been summarised into three categories that are discussed: a) physical planning and zoning, b) promoting pollution control, and c) cleaner production. It contains examples where strategies have been implemented and the advantages and disadvantages.

*a) Physical planning and zoning*

— Restricting the location of new industries; vicinity to residential zones or other sensitive areas should be prohibited;

— Relocation of existing industries away from residential zones if economically viable;

— Controlling of emissions in sensitive areas, for example by implementing special control areas or smoke-free zones;

— Area planning - based on emission assessments and air quality objectives, cities can determine what kind of industrial activities and pollution control equipment is required per area (including sensitive areas as well as industrial and commercial areas).

— Environmental Impact Assessment (EIA) for new major industries in order to assess the environmental impact – in this case, the potential air pollution - of new activities and develop recommendations for improvement in terms of location choice, type of industry and pollution equipment. It is desirable to make the EIA compulsory. The report 'An approach to environmental impact assessment for projects affecting the coastal and marine environment'is a short easy-to-read report explaining the general procedures for EIA. Although it has been developed for coastal water systems, the approach could also be used in the field of air quality.

*b) Promoting pollution control in factories*

— Setting priorities by focusing on pollution control devices for the most serious polluting factories

— Using best available techniques

It is strongly recommended to focus urban air quality policy on the implementation of best available/affordable techniques for specific industrial processes. The advantage is that local governments can make an action plan for the implementation of best available and/or affordable techniques in cooperation with the industries themselves. Experiences have illustrated that this often results in a realistic action plan generating commitment from all of those involved. Furthermore, it is easier to monitor whether industries have installed pollution control devices than to control their exact emissions.

— Compulsory notification of accidents

— Licensing of specified strongly polluting processes

— Compulsory emission limits

Many developing countries have set emission standards for specific types of industries. However, experience shows that enforcement is often weak; it is thus recommended that this enforcement strategy be combined with the best available/affordable techniques.

— Setting strict fines for exceeding emission limits

*c) Promoting cleaner production (prevention solutions rather then end-of-pipe remedies)*

Cleaner production means increasing the efficiency of industrial processes and the use of products to prevent the pollution of air, water, and land; reducing wastes at their source; and minimising risks to people and the environment. Cleaner production is a way to achieve both environmental protection and economic benefits. By better managing the production process, one can save energy and materials.

In the case of air quality management, the main cleaner production successes have been achieved by improving fuel quality. Examples include restricting the sulphur content of fuels and encouraging the use of cleaner types of fuels (such as natural gas). Although cleaner production is the most sustainable solution, one should realise that end-ofpipe solutions are still necessary to address industrial air pollution. Many end-of-pipe techniques achieve reductions of over 90% in air emissions, something that could hardly be achieved with cleaner production measures alone.

# 13

# Emissions Trading

Emissions trading (or emission trading) is an administrative approach used to control pollution by providing economic incentives for achieving reductions in the emissions of pollutants. It is sometimes called cap and trade.

Due to emissions trading, coal might become less competitive as a fuel. A central authority sets a limit or cap on the amount of a pollutant that can be emitted. Companies or other groups are issued emission permits and are required to hold an equivalent number of allowances (or credits) which represent the right to emit a specific amount. The total amount of allowances and credits cannot exceed the cap, limiting total emissions to that level. Companies that need to increase their emission allowance must buy credits from those who pollute less. The transfer of allowances is referred to as a trade. In effect, the buyer is paying a charge for polluting, while the seller is being rewarded for having reduced emissions by more than was needed. Thus, in theory, those that can easily reduce emissions most cheaply will do so, achieving the pollution reduction at the lowest possible cost to society.

There are active trading programs in several pollutants. For greenhouse gases the largest is the

European Union Emission Trading Scheme. In the United States there is a national market to reduce acid rain and several regional markets in nitrous oxide. Markets for other pollutants tend to be smaller and more localised.

Carbon trading is sometimes seen as a better approach than a direct carbon tax or direct regulation. By solely aiming at the cap it avoids the consequences and compromises that often accompany other methods. It can be cheaper, and politically preferable for existing industries because the initial allocation of allowances is often allocated with a grandfathering provision where rights are issued in proportion to historical emissions. In addition, most of the money in the system is spent on environmental activities, and the investment directed at sustainable projects that earn credits in the developing world can contribute to the Millennium Development Goals. Critics of emissions trading point to problems of complexity, monitoring, enforcement, and sometimes dispute the initial allocation methods and cap.

The overall goal of an emissions trading plan is to reduce emissions. The cap is usually lowered over time - aiming towards a national emissions reduction target. In other systems a portion of all traded credits must be retired, causing a net reduction in emissions each time a trade occurs. In many cap and trade systems, organisations which do not pollute may also participate, thus environmental groups can purchase and retire allowances or credits and hence drive up the price of the remainder according to the law of demand. Corporations can also prematurely retire allowances by donating them to a nonprofit entity and then be eligible for a tax deduction.

Because emissions trading uses markets to determine how to deal with the problem of pollution, it is often touted as an example of effective free market environmentalism. While the cap is usually set by a political process, individual companies are free to choose how or if they will reduce their emissions. In theory, firms will choose the least-costly way to comply with the pollution regulation, creating incentives that reduce the cost of achieving a pollution reduction goal.

## Historical Outlook

The efficacy of, what later was to be called, the "Cap and Trade" approach to air pollution abatement was first demonstrated in a series of micro-economic computer simulation studies between 1967 and 1970 for the National Air Pollution Control Administration by Ellison Burton and William Sanjour. These studies used mathematical models of several cities and their emission sources in order to compare the cost and effectiveness of various control strategies. For each abatement strategy comparison was made with the "least cost solution" produced by a computer optimisation program which finds the least costly combination of source reductions to achieve a given abatement goal. In each case it was found that the least cost solution was dramatically less costly for the same level of pollution produced by any conventional abatement strategy. This led to the concept of "Cap and Trade" as a means of achieving the "least cost solution" for a given level of abatement.

The development of emissions trading over the course of its history can be divided into four phases:

1. *Gestation:* Theoretical articulation of the instrument (by Coase, Dales, Montgomery etc) and, independent of the former, tinkering with "flexible regulation" at the US Environmental Protection Agency
2. *Proof of Principle:* First developments towards trading of emission certificates based on the "offset-mechanism" taken up in Clean Air Act in 1977.

3. *Prototype:* Launching of a first "cap and trade" system as part of US Acid Rain Program, officially announced as a paradigm shift in environmental policy, as prepared by "Project 88", a network building effort to bring together environmental and industrial interests in the US
4. *Regime formation:* branching out from US clean air policy to global climate policy, and from there to the European Union, along with the expectation of an emerging global carbon market and the formation of the "carbon industry".

**Cap and Trade Versus Baseline and Credit**

The textbook emissions trading program can be called a "cap and trade" approach in which an aggregate cap on all sources is established and these sources are then allowed to trade amongst themselves to determine which sources actually emit the total pollution load. An alternative approach with important differences is a baseline and credit program. In a baseline and credit program a set of polluters that are not under an aggregate cap can create credits by reducing their emissions below a baseline level of emissions. These credits can be purchased by polluters that do have a regulatory limit. Many of the criticisms of trading in general are targeted at baseline and credit programs rather than cap type programs.

**Economics of International Emissions Trading**

It's possible for a country to reduce emissions using a Command-Control approach, such as regulation, direct and indirect taxes. But that approach is more costly for some countries than for others. That's because the Marginal Abatement Cost (MAC) — the cost of eliminating an additional unit of pollution — differs by country. It might cost China \$2 to eliminate a ton of $CO_2$, but it would probably cost Sweden or the U.S. much more. International emissions-trading markets were created precisely to exploit differing MACs.

**Applying the Economic Theory**

The nature of the pollutant plays a very important role when policy-makers decide which framework should be used to control pollution. $CO_2$ acts globally, thus its impact on the environment is generally similar wherever in the globe it is released. So the location of the originator of the emissions does not really matter from an environmental standpoint.

The policy framework should be different for regional pollutants (e.g. $SO_2$ and $NO_X$, and also Mercury) because the impact exerted by these pollutants may not be the same in all locations. The same amount of a regional pollutant can exert a very high impact in some locations and a low impact in other locations, so it does actually matter where the pollutant is released. This is known as the 'Hot Spot' problem.

A Lagrange framework is commonly used to determine the least cost of achieving an objective, in this case the total reduction in emissions required in a year. In some cases it is possible to use the Lagrange optimisation framework to determine the required reductions for each country (based on their MAC) so that the total cost of reduction is minimised. In such a scenario, the Lagrange Multiplier represents the market allowance price (P) of a pollutant, such as the current market allowance price of emissions in Europe and the USA.

All countries face the market allowance price as existent in the market that day, so they are able to make individual decisions that would maximise their profit while at the same time achieving regulatory compliance.

This is also another version of the Equi-Marginal Principle, commonly used in economics to choose the most economically efficient decision.

***Prices versus quantities, and the safety valve***

There has been longstanding debate on the relative merits of price versus quantity instruments to achieve emission reductions.

- An emission cap and permit trading system is a quantity instrument because it fixes the overall emission level (quantity) and allows the price to vary. One problem with the cap and trade system is the uncertainty of the cost of compliance as the price of a permit is not known in advance and will vary over time according to market conditions.
- In contrast, an emission tax is a price instrument because it fixes the price while the emission level is allowed to vary according to economic activity. A major drawback of an emission tax is that the environmental outcome (e.g. a limit on the amount of emissions) is not guaranteed.

The best choice depends on the sensitivity of the costs of emission reduction, compared to the sensitivity of the benefits (i.e., climate damages avoided by a reduction) when the level of emission control is varied.

Because there is high uncertainty in the compliance costs of firms, some argue that the optimum choice is the price mechanism.

However, some scientists have warned of a threshold in atmospheric concentrations of carbon dioxide beyond which a run-away warming effect could take place, with a large possibility of causing irreversible damages. If this is a conceivable risk then a quantity instrument could be a better choice because the quantity of emissions may be capped with a higher degree of certainty. However, this may not be true if this risk exists but cannot be attached to a known level of GHG concentration or a known emission pathway.

A third option, known as a safety valve, is a hybrid of the price and quantity instruments. The system is essentially an emission cap and tradeable permit system but the maximum (or minimum) permit price is capped. Emitters have the choice of either obtaining permits in the marketplace or purchasing them from the government at a specified trigger price (which could be adjusted over time). The system is sometimes recommended as a way of overcoming the fundamental disadvantages of both systems by giving governments the flexibility to adjust the system as new information comes to light. It can be shown that by setting the trigger price high enough, or the number of permits low enough, the safety valve can be used to mimic either a pure quantity or pure price mechanism.

All three methods are being used as policy instruments to control greenhouse gas emissions: the EU-ETS is a quantity system using the cap and trading system to meet targets set by National Allocation Plans, the UK's Climate Change Levy is a price system using a direct carbon tax, while China uses the $CO_2$ market price for funding of its Clean Development Mechanism projects, but imposes a safety valve of a minimum price per tonne of $CO_2$.

## Trading Systems

### Kyoto Protocol

The Kyoto Protocol is a 1997 international treaty which came into force in 2005, which binds most developed nations to a cap and trade system for the six major greenhouse gases. Emission quotas were agreed by each participating country, with the intention of reducing their

After voluntary trials in the UK and Denmark, Phase I commenced operation in January 2005 with all 15 (now 25 of the 27) member states of the European Union participating. The program caps the amount of carbon dioxide that can be emitted from large installations, such as power plants and carbon intensive factories and covers almost half of the EU's Carbon Dioxide emissions. Phase I permits participants to trade amongst themselves and in validated credits from the developing world through Kyoto's Clean Development Mechanism.

Whilst the first phase has received much criticism due to oversupply of allowances and the distribution method of allowances, Phase II links the ETS to other countries participating in the Kyoto trading system. The European Commission has been tough on Member States' Plans for Phase II, dismissing many of them as being too loose again. In addition, the first phase has established a strong carbon market. Compliance was high in 2006, increasing confidence in the scheme, although the value of allowances dropped when the national caps were met.

All EU member states have ratified the Kyoto Protocol, and so the second phase of the EU ETS has been designed to support the Kyoto mechanisms and compliance period. Thus any organisation trading through the ETS should also meet the international trading obligations under Kyoto.

*New Zealand*

The New Zealand Government introduced a bill for emissions trading schemes before a select committee. Various reports by a range of groups support the scheme but differ in opinion as to how it should be implemented. An interesting feature of the New Zealand Emissions Trading Scheme is that it includes forest carbon and creates deforestation liabilities for landowners.

The emissions trading bill passed into law on 10 September 2008. On November 16 2008 the newly formed National-led government announced that it would delay implementation of the ETS pending a full review of climate change policy.

***United States***

An early example of an emission trading system has been the $SO_2$ trading system under the framework of the Acid Rain Program of the 1990 Clean Air Act in the U.S. Under the program, which is essentially a cap-and-trade emissions trading system, $SO_2$ emissions are expected to be reduced by 50 percent from 1980 levels by 2010. Some experts argue that the "cap and trade" system of $SO_2$ emissions reduction has reduced the cost of controlling acid rain by as much as 80 percent versus source-by-source reduction.

In 1997, the State of Illinois adopted a trading program for volatile organic compounds in most of the Chicago area, called the Emissions Reduction Market System. Beginning in 2000, over 100 major sources of pollution in eight Illinois counties began trading pollution credits.

In 2003, New York State proposed and attained commitments from nine Northeast states to form a cap and trade carbon dioxide emissions program for power generators, called the Regional Greenhouse Gas Initiative (RGGI). This program is due to launch on January 1, 2009 with the aim to reduce the carbon "budget" of each state's electricity generation sector to 10 percent below their 2009 allowances by 2018.

Also in 2003, U.S. corporations were able to trade $CO_2$ emission allowances on the Chicago Climate Exchange under a voluntary scheme. In August 2007, the Exchange announced a mechanism to create emission offsets for projects within the United States that cleanly destroy ozone-depleting substances.

In 2007, the California Legislature passed the California Global Warming Solutions Act, AB-32, which was signed into law by Governor Arnold Schwarzenegger. Thus far, flexible mechanisms in the form of project based offsets have been suggested for five main project types. A carbon project would create offsets by showing that it has reduced carbon dioxide and equivalent gases. The project types include: manure management, forestry, building energy, SF6, and landfill gas capture. California is also one of seven states and three Canadian province that have joined together to create the Western Climate Initiative, which has recommended the creation of a regional greenhouse gas control and offset trading environment.

In a talk recorded for YouTube President Elect Barack Obama on Nov 17 2008 clarified that the US will enter a cap and trade system to limit Global Warming.

*Renewable energy certificates*

Renewable Energy Certificates, or "green tags", are transferable rights for renewable energy within some American states. A renewable energy provider gets issued one green tag for each 1,000 KWh of energy it produces. The energy is sold into the electrical grid, and the certificates can be sold on the open market for additional profit. They are purchased by firms or individuals in order to identify a portion of their energy with renewable sources and are voluntary.

They are typically used like an offsetting scheme or to show corporate responsibility, although their issuance is unregulated, with no national registry to ensure there is no double-counting. However, it is one way that an organisation could purchase its energy from a local provider who uses fossil fuels, but back it with a certificate that supports a specific wind or hydro power project.

## Carbon Market

Carbon emissions trading is emissions trading specifically for carbon dioxide (calculated in tonnes of carbon dioxide equivalent or $tCO_2e$) and currently makes up the bulk of emissions trading. It is one of the ways countries can meet their obligations under the Kyoto Protocol to reduce carbon emissions and thereby mitigate global warming.

### Market trend

Carbon emissions trading has been steadily increasing in recent years. According to the World Bank's Carbon Finance Unit, 374 million metric tonnes of carbon dioxide equivalent ($tCO_2e$) were exchanged through projects in 2005, a 240% increase relative to 2004 (110 $mtCO_2e$) which was itself a 41% increase relative to 2003 (78 $mtCO_2e$).

In terms of dollars, the World Bank has estimated that the size of the carbon market was 11 billion USD in 2005, 30 billion USD in 2006, and 64 billion in 2007. The Marrakesh Accords of the Kyoto protocol defined the international trading mechanisms and registries needed to support trading between countries, with allowance trading now occurring between European countries and Asian countries. However, while the USA as a nation did not ratify the Protocol, many of its states are now developing cap-and-trade systems and are looking at ways to link their emissions trading systems together, nationally and internationally, to seek out the lowest costs and improve liquidity of the market. However, these states also wish to preserve their individual integrity and unique features. For example, in contrast to the other Kyoto-compliant systems, some states propose other types of greenhouse gas sources, different measurement methods, setting a maximum on the price of allowances, or restricting access to CDM projects. Creating instruments that are not truly fungible would introduce instability and make pricing difficult. Various proposals are being investigated to see how these systems might be linked across markets, with the International Carbon Action Partnership (ICAP) as an international body to help co-ordinate this.

**Business reaction**

With the creation of a market for mandatory trading of carbon dioxide emissions within the Kyoto Protocol, the London financial marketplace has established itself as the center of the carbon finance market, and is expected to have grown into a market valued at $60 billion in 2007. The voluntary offset market, by comparison, is projected to grow to about $4bn by 2010.

23 multinational corporations came together in the G8 Climate Change Roundtable, a business group formed at the January 2005 World Economic Forum. The group included Ford, Toyota, British Airways, BP and Unilever. On 9 June 2005 the Group published a statement stating that there was a need to act on climate change and stressing the importance of market-based solutions. It called on governments to establish "clear, transparent, and consistent price signals" through "creation of a long-term policy framework" that would include all major producers of greenhouse gases. By December 2007 this had grown to encompass 150 global businesses.

Business in the UK have come out strongly in support of emissions trading as a key tool to mitigate climate change, supported by NGOs. However, not all businesses favor a trading approach. On December 11, 2008, Rex Tillerson, the CEO of Exxonmobil, said a carbon tax is "a more direct, more transparent and more effective approach" than a cap and trade program, which he said, "inevitably introduces unnecessary cost and complexity." He also said that he hoped that the revenues from a carbon tax would be used to lower other taxes so as to be revenue neutral.

**Measuring, Reporting, Verification (MRV) and Enforcement**

Meaningful emission reductions within a trading system can only occur if they can be measured at the level of operator or installation and reported to a regulator. For greenhouse gases all trading countries maintain an inventory of emissions at national and installation level; in addition, the trading groups within North America maintain inventories at the state level through The Climate Registry. For trading between regions these inventories must be consistent, with equivalent units and measurement techniques.

In some industrial processes emissions can be physically measured by inserting sensors and flowmeters in chimneys and stacks, but many types of activity rely on theoretical calculations for measurement. Depending on local legislation, these measurements may require additional checks and verification by government or third party auditors, prior or post submission to the local regulator.

Another critical part is enforcement. Without effective MRV and enforcement the value of allowances are diminished. Enforcement can be done using several means, including fines or sanctioning those that have exceeded their allowances. Concerns include the cost of MRV and enforcement and the risk that facilities may be tempted to mislead rather than make real reductions or make up their shortfall by purchasing allowances or offsets from another entity. The net effect of a corrupt reporting system or poorly managed or financed regulator may be a discount on emission costs, and a (hidden) increase in actual emissions.

**Limitations**

There are critics of the methods, mainly environmental justice NGOs and movements, who see carbon trading as a proliferation of the free market into public spaces and environmental policy-making. They level accusations of failures in accounting, dubious science and the destructive impacts of projects upon local peoples and environments as reasons why trading pollution allowances should be avoided. In its place they advocate making reductions at the source of pollution and energy policies that are justice-based and community-driven. Most of the criticisms have been focused on the carbon market created through investment in Kyoto Mechanisms. Criticism of 'cap and trade' emissions trading has generally been more limited to lack of credibility in the first phase of the EU ETS.

Critics argue that emissions trading does little to solve pollution problems overall, as groups that do not pollute sell their conservation to the highest bidder. Overall reductions would need to come from a sufficient and challenging reduction of allowances available in the system.

Regulatory agencies run the risk of issuing too many emission credits, diluting the effectiveness of regulation, and practically removing the cap. In this case, instead of any net reduction in carbon dioxide emissions, beneficiaries of emissions trading simply do more of the polluting activity. The National Allocation Plans by member governments of the European Union Emission Trading Scheme were criticised for this when it became apparent that actual emissions would be less than the government-issued carbon allowances at the end of Phase I of the scheme.

They have also been criticised for the practice of grandfathering, where polluters are given free allowances by governments, instead of being made to pay for them. Critics instead advocate for auctioning the credits. The proceeds could be used for research and development of sustainable technology.

Critics of carbon trading, such as Carbon Trade Watch, argue that it places disproportionate emphasis on individual lifestyles and carbon footprints, distracting attention from the wider, systemic changes and collective political action that needs to be taken to tackle climate change. Groups such as the Corner House (organisation) have argued that the market will choose the easiest means to save a given quantity of carbon in the short term, which may be different to the pathway required to obtain sustained and sizable reductions over a longer period, and so a market led approach is likely to reinforce technological lock-in. For instance small cuts may often be achieved cheaply through investment in making a technology more efficient, where larger cuts would require scrapping the technology and using a different one. They also argue that emissions trading is undermining alternative approaches to pollution control with which it does not combine well, and so the overall effect it is having is to actually stall significant change to less polluting technologies.

Although many economists advocate carbon trading schemes, some others favor a carbon tax instead. Possible problems with cap and trade systems include:

- Trading may be a more complicated means of achieving the same objective
- Permit prices may be unstable and therefore unpredictable
- Some cap and trade systems pass the quota rent to business though grandfathering (a certain number of

credits are given away for free rather than auctioned)

- Cap and trade systems could become the basis for international trade in the quota rent resulting in very large transfers across frontiers
- Cap and trade systems are seen to generate more corruption than a tax system
- The administration and legal costs of cap and trade systems are higher than with a tax
- A cap and trade system is seen to be impractical at level of individual household emissions

The problem of unstable prices can be resolved, to some degree, by the creation of forward markets in caps. Nevertheless, it is easier to make a tax predictable than the price of a cap. However, the corresponding uncertainty under a tax is the level of emissions reductions achieved.

The Financial Times wrote an article on cap and trade systems that argues that "Carbon markets create a muddle" and "...leave much room for unverifiable manipulation".

More recent criticism of emissions trading regarding implementation is that old growth forests are being cleared and replaced with fast-growing vegetation, to the detriment of the local communities.

Recent proposals for alternative schemes that seek to avoid the problems of Cap and Trade schemes include Cap and Share, which at May 2008 is being actively considered by the Irish Parliament, and the Sky Trust schemes.

## Action against Climate Change

The EU emissions trading scheme (ETS) is based on a recognition that creating a price for carbon through the establishment of a liquid market for emission reductions offers the most cost-effective way for EU Member States to meet their Kyoto obligations and move towards the low-carbon economy of the future.

The scheme should allow the EU to achieve its Kyoto target at a cost of between EUR 2.9 billion and EUR 3.7 billion annually. This is less than 0.1 % of the EU's GDP. Without the scheme, compliance costs could reach up to EUR 6.8 billion a year.

The ETS has been established through binding legislation (2) proposed by the European Commission and approved by the EU Member States and the European Parliament. The scheme is based on six fundamental principles.

— It is a 'cap-and-trade' system.
— Its initial focus is on $CO_2$ from big industrial emitters.
— Implementation is taking place in phases, with periodic reviews and opportunities for expansion to other gases and sectors.
— Allocation plans for emission allowances are decided periodically.
— It includes a strong compliance framework.
— The market is EU-wide but taps emission reduction opportunities in the rest of the world through the use of the CDM and JI, and provides for links with compatible schemes in third countries.

**Scheme Covers**

While emissions trading has the potential to involve many sectors of the economy and all the greenhouse gases controlled by the Kyoto Protocol ($CO_2$, methane, nitrous oxide, hydrofluorocarbons, perfluorocarbons and sulphur hexafluoride), the scope of the ETS is intentionally limited during its initial phase while experience of emissions trading is built up.

Consequently, during the first trading period from 2005 to 2007, the ETS covers only $CO_2$ emissions from large emitters in the power and heat generation industry and in selected energy-intensive industrial sectors: combustion plants, oil refineries, coke ovens, iron and steel plants and factories making cement, glass, lime, bricks, ceramics, pulp and paper. A size threshold based on production capacity or output determines which plants in these sectors are included in the scheme.

Even with this limited scope, close to 11 500 installations in the 25 Member States are covered, accounting for around 45 % of the EU's total $CO_2$ emissions or about 30 % of its overall greenhouse gas emissions. Towards the end of 2005, the European Commission also started work on including emissions from aviation in the ETS. By mid-2006, the Commission is due to present a report reviewing the functioning of the scheme. The review will allow fine-tuning it in the light of the experience gained and to consider whether it should be extended to additional sectors and more greenhouse gases.

**Emission Allowances**

At the heart of the ETS is the common trading 'currency' of emission allowances. One allowance represents the right to emit one tonne of $CO_2$. Member States have drawn up national allocation plans for 2005–07 which give each installation in the scheme a certain number of allowances free of charge, thus allowing it to emit the corresponding amount of $CO_2$ without any cost. Decisions on the allocations are made public.

The limit or 'cap' on the number of allowances allocated creates the scarcity needed for a trading market to emerge. Companies that keep their emissions below the level of their allowances are able to sell their excess allowances at a price determined by supply and demand at that time. Those facing difficulty in remaining within their emissions limit have a choice between taking measures to reduce their emissions, such as investing in more efficient technology or using a less carbon-intensive energy source, buying the extra allowances they need at the market rate, or a combination of the two, whichever is cheapest. This ensures that emissions are reduced in the most cost-effective way. Most allowances are allocated to installations free of charge — at least 95 % during the initial phase and at least 90 % in the second phase from 2008 to 2012. Though only plants covered by the scheme are given allowances, anyone else — individuals, institutions, non-governmental organisations or whoever — is free to buy and sell in the market in the same way as companies.

**National Allocation Plans**

Member States' national allocation plans (NAPs) have to be based on objective and transparent criteria, including a set of common rules that are laid down in the legislative framework establishing the ETS. The most important of these rules are listed below.

- An allocation plan has to reflect a Member State's Kyoto target as well as its actual and projected progress towards meeting it. The total quantity of allowances allocated is key in this regard. Allocating too many allowances would mean that greater efforts to cut emissions would have to be taken in economic sectors not covered by the scheme, in potentially less cost-effective ways than trading.
- Allocations to installations must take account of their potential for reducing emissions from each of their

activities, and must not be higher than the installations are likely to need.

— Where Member States intend to use JI and CDM credits to help them reach their national emission target — thereby giving their companies more scope to emit — these plans must be substantiated, for example through budgetary provisions.

The European Commission has issued specific guidance on how these rules are to be applied by Member States. The Commission assesses NAPs on the basis of these rules, as well as EU rules on State aid and competition, and has the power to require changes or even to reject a plan altogether. Once it approves a plan, the total quantity of allowances cannot be changed; nor can the number of allowances per installation following the final allocation by the Member State.

**Ensuring Compliance**

Appropriately for a market-based instrument that makes it possible to put a price on carbon, the ETS incorporates a robust framework of measures to ensure compliance that also gives a central role to economic incentives.

After each calendar year, installations must surrender a number of allowances equivalent to their verified $CO_2$ emissions in that year. These allowances are then cancelled so they cannot be used again. Those installations with allowances left over can sell them or save them for next year. Those that have not produced enough allowances to cover their emissions will have to pay a dissuasive fine for each excess tonne emitted. In the initial phase the penalty is EUR 40 per tonne, but from 2008 it will rise to EUR 100. Operators also have to obtain allowances to make up the shortfall in the following year, and they will be 'named and shamed' by having their names published. Member States have also had to lay down dissuasive penalties for any infringements of the ETS rules at national level.

### Monitoring and Reporting Emissions

Each installation in the ETS must have a permit from its competent authority for its emissions of all six greenhouse gases controlled by the Kyoto Protocol. A condition for granting the permit is that the operator is capable of monitoring and reporting the plant's emissions. A permit is different from the allowances: the permit sets out the emissions monitoring and reporting requirements for an installation, whereas allowances are the scheme's tradable unit.

Installations must report their $CO_2$ emissions after each calendar year. The European Commission has issued a set of monitoring and reporting guidelines (4) to be followed. Installations' reports have to be checked by an independent verifier on the basis of criteria set out in the ETS legislation, and are made public. Operators whose emission reports for the previous year are not verified as satisfactory will not be allowed to sell allowances until a revised report is approved by a verifier.

### Transaction Registries

Allowances are not printed but held in accounts in electronic registries set up by Member States. The European Commission has set out specific legislation for a standardised and secured system of registries based on UN data exchange standards to track the issue, holding, transfer and cancellation of allowances. Provisions on the tracking and use of credits from JI and CDM projects in the EU scheme are also included. The registries system is similar to a banking system which keeps track of the ownership of money in accounts but does not look into the deals that lead to money changing hands. The system of registries is overseen by a central administrator at EU level who, through an independent transaction log, checks each transaction for any irregularities. Any irregularities detected prevent a transaction from being completed until they have been remedied. The EU registries system is being integrated with the international registries system used under the Kyoto Protocol.

**Trading in Practice**

The legal framework of the ETS does not lay down how and where trading in allowances should take place. Companies and other participants in the market trade directly with each other or buy and sell via a broker, exchange or any other type of market intermediary that has sprung up to take advantage of this significant new market.

The price of allowances is determined by supply and demand as in any other market. Trading on a forward basis began as early as 2003. It has significantly increased since the emissions trading scheme officially kicked off in January 2005, with spot trading gaining ground as Member States set up their national registries and issued allowances into the accounts of the companies covered by the scheme.

Traded volumes have been increasing accordingly. During the first six months of 2005, the market saw transactions of more than 90 million EU allowances (65.6 million brokered, 10.2 million traded on exchanges, and an estimated 15 million traded bilaterally). This corresponds to an estimated financial volume of EUR 1.37 billion. With this, the ETS has established itself as the engine of the global carbon market, which is emerging as a powerful tool to combat climate change. The scheme is generating important knowledge for the successful design and operation of carbon trading schemes elsewhere and international emissions trading to start in

2008 under the Kyoto Protocol, and it has given strong impetus to CDM and JI projects. In the EU, businesses are learning to operate in a carbon-constrained environment and to develop the best strategies to reduce their greenhouse gas emissions costeffectively. Joint implementation and the clean development mechanism enable developed countries that have binding emission reduction or limitation targets under the Kyoto Protocol to undertake emission-saving investments in third countries and credit these savings towards their own emission target.

The CDM covers projects in countries without an emission target under the protocol, i.e. developing nations. Reductions since 2000 are potentially eligible to receive credits called 'certified emission reductions', or CERs. JI applies to projects in countries that have agreed to an emission target —other industrialised countries and countries with economies in transition — and will yield credits known as 'emission reduction units', or ERUs, once the first Kyoto commitment period starts in 2008. The EU scheme is the first in the world that recognises most of these credits as equivalent to emission allowances (1 EUA = 1 CER = 1 ERU) and allows them to be traded under the scheme. Credits from nuclear facilities and land use, land-use change and forestry activities are not accepted. Thus, the launch of the ETS three years before the start of the protocol's first commitment period has given certainty to investors in the rapidly emerging market for JI and CDM projects. This has encouraged more investment in such projects and consequently promotes the transfer of environmentally sound technologies that help the host countries meet their sustainable development goals.

For EU companies covered by the scheme, the recognition of JI and CDM credits has increased the range of options available for meeting their emission targets, improved the liquidity of the market and potentially lowered the price of allowances, thus further reducing their compliance costs. Companies are not the only ones looking for emission reduction credits through JI and CDM. Member States intend to use such credits themselves to help meet their emission target under the protocol. In their national allocation plans, they have indicated that they intend to procure 500–600 million tonnes of $CO_2$ credits for the period 2008–12. Since the majority of JI and CDM projects tend to generate emission reductions averaging between 500 000 and 1 million tonnes of $CO_2$, EU countries' demand for emission credits can only be satisfied through a great number of such projects. As 2008 draws nearer, EU Member States are actively seeking JI and CDM projects and a number of project contracts have already been signed.

With this strong demand for emission credits building up rapidly, major European banks and other financial institutions both in the private and public sector have become active in providing finance for prospective emission reduction projects.

But the use of the Kyoto mechanisms within the EU will be only supplemental to domestic action to limit or reduce emissions, as agreed by UNFCCC parties at Marrakesh in 2001. Linking the EU ETS with trading schemes in partner countries The ETS is open to linking with compatible greenhouse gas emission trading schemes in other countries that have ratified the Kyoto Protocol. It is foreseen that each side would agree to recognise allowances issued by the other, thereby expanding the market for trading.

The EU is discussing Norway's participation in the ETS and preliminary discussions on cooperation have taken place with a number of other countries. The EU is also encouraged by moves to create an emissions trading scheme for $CO_2$ among initially nine north-eastern US states.

Shaping the future debate by sharing experience with partners During the design of this innovative scheme, EU officials have exchanged information with many experts about existing trading schemes for other pollutants, particularly in the USA. Now that the EU scheme has started, systematic 'learning by doing' will be crucial for its successful further development. Independent monitoring and evaluation is accompanying its implementation. These processes will yield invaluable information which the EU will want to share with all interested parties and stakeholders.

# Bibliography


Nyberg F, Gustavsson P, Jarup L et al., "Urban air pollution and lung cancer in Stockholm", *Epidemiology,* 2000.

O'Neill M, Jerrett M, Kawachi I et al., "Health, wealth and air pollution", *Environmental Health Perspectives,* 2003.

Ostro B, Chestnut L., "Assessing the health benefits of reducing particulate matter air pollution in the United States", *Environmental Research,* 1998.

Ostro B, Chestnut L, Vichit-Vadakan N, Laixuthai A., "The impact of particulate matter on daily mortality in Bangkok", Thailand. *Journal of the Air and Waste Management Association,* 1999.

Ostro B, Hurley S, Lipsett MJ., "Air pollution and daily mortality in the Coachella Valley", California: a study of PM10 in an area dominated by coarse particles, *Environmental Research,* 1999.

Ostro B.D, Sanchez JM, Aranda C, Eskeland G.S., "Air pollution and mortality: results from a study of Santiago", Chile. *Journal of Exposure Analysis and Environmental Epidemiology,* 1996.

Ozkaynak H., "Exposure assessment", In: *Air pollution and health.* Holgate ST, Samet JM, Koren HS, Maynard R, eds. Academic Press, London, 1999.

Pachauri, R.K.; Sridharan, P.V., *Looking Back* to *Think Ahead: growth with resource enhancement of environment and nature,* New Delhi: Tata Energy Research Institute, 1998.

Pope C.A III, Dockery D.W., "Epidemiology of particle effects", In: *Air pollution and health.* Holgate ST, Samet JM, Koren HS, Maynard R, eds. Academic Press, London, 1999.

Pope CA III, Thun MJ, Namboodiri MM et al., "Particulate air pollution as a predictor of mortality in a prospective study of US adults", *American Journal of Respiratory and Critical Care Medicine,* 1995.

Qian Z, Chapman RS, Tian Q, Chen Y, Lioy PJ, Zhang J., "Effects of air pollution on children's respiratory health in three Chinese cities", *Archives of Environmental Health,* 2000.

Qian Z, Zhang J, Wei F, Wilson WE, Chapman R.S., "Long term ambient air pollution levels in four Chinese cities: inter-city and intra-city concentration gradients for epidemiological studies", *Journal of Exposure Analysis and Environmental Epidemiology,* 2001.

Reddy KS, Yusuf S., "Emerging epidemic of cardiovascular disease in developing countries", *Circulation,* 1998.

Roemer W, Clench-Aas J, Englert N et al., "Inhomogeneity in response to air pollution in European children (PEACE project)", *Occupational and Environmental Medicine,* 1999.

Romieu I, Samet JM, Smith KR, Bruce N., "Outdoor air pollution and acute respiratory infections among children in developing countries", *Journal of Occupational and Environmental Medicine,* 2002.

Saldiva PH, Pope CA, III, Schwartz J et al., "Air pollution and mortality in elderly people: a time-series study in São Paulo, Brazil", *Archives of Environmental Health,* 1995.

World Bank, *World Development report: investing in health,* Washington, DC: The World Bank, Oxford University Press, 1993.

World Bank, *India's Environment: taking stock of plans, programs and priorities,* South Asia Regional Office, The World Bank, 1996.

WHO/UNEP, *Urban air pollution in megacities of the World,* Oxford: Basil Blackwell, World Health Organisation/United Nations Environment Programme, 1992.

# Index